HENRY PURCELL

R

Oxford Studies of Composers

Henry Purcell

PETER HOLMAN

Oxford New York
OXFORD UNIVERSITY PRESS
1994

...eet, Oxford OX2 6DP
...k
... Bombay
...Salaam Delhi
...ul Karachi
...rid Melbourne
Mexico City Nairobi Paris Singapore
Taipei Tokyo Toronto

and associated companies in
Berlin Ibadan

Oxford is a trade mark of Oxford University Press

Published in the United States
by Oxford University Press Inc., New York

British Library Cataloguing in Publication Data
Data available
ISBN 0-19-816340-1 (acid-free paper)
ISBN 0-19-816341-X (pbk.: acid-free paper)

Library of Congress Cataloging in Publication Data
Holman, Peter.
Henry Purcell / Peter Holman.
p. cm.—(Oxford studies of composers)
Includes bibliographical references (p.) and indexes.
1. Purcell, Henry, 1659–1695—Criticism and interpretation.
I. Title. II. Series.
ML410.P93H63 1995 780'.92—dc20 94-31870
ISBN 0-19-816340-1 (acid-free paper)
ISBN 0-19-816341-X (pbk.: acid-free paper)

Set by Hope Services (Abingdon) Ltd.
Printed in Great Britain
on acid-free paper by
Biddles Ltd., Guildford & King's Lynn

for Tricia

PREFACE

THESE are interesting times for lovers of Henry Purcell's music. The commercial bandwagon of the 1995 tercentenary has begun to roll, propelled by ambitious plans for performances, recordings, conferences, exhibitions, films, television programmes, and media events. The complete edition of Purcell's music, started by the Purcell Society in 1878, and under revision since the 1960s, will soon be complete. At the present rate of progress it will not be long before virtually all his music is available on CD.

Purcell scholarship, too, is on the move after a fallow period. Thanks to Andrew Ashbee's splendid series of *Records of English Court Music*, we now have a much clearer idea of musical life at the Restoration court. Curtis Price, Andrew Pinnock, Bruce Wood, and others have thrown new light on Purcell's theatre music, re-evaluating the sources, and drawing on literary and theatre scholarship to a greater extent than before. As a result, a number of cherished notions have been challenged: we are no longer so sure that *Dido and Aeneas* was written in 1689 for Josiah Priest's girls' school in Chelsea, and that it is an allegory of the Glorious Revolution. There is renewed interest in Purcell's autographs, fuelled by the discovery of two new examples in the last few months: a volume of keyboard music copied partly by Purcell and partly by his Italian colleague Giovanni Battista Draghi (see Ch. 3), and a single sheet containing an anthem by Daniel Roseingrave, the music copied by Purcell, the words added by the composer (see Ch. 1).*

The prospective writer on Purcell can only view all this with a mixture of elation and alarm. I am only too aware that new research over the next few years will probably make parts of this book out of date. Purcell scholars are only now beginning to investigate the sources of his music in detail, looking at paper types, watermarks, the evidence of rastra, and so on. Studies of this sort have revolutionized our knowledge of the music of Bach and Mozart, and there is every reason to think that they will do

* Note: As this book goes to press news comes of a third new Purcell autograph, an organ part for John Blow's anthem 'God is our hope and strength', discovered by Robert Thompson in GB–Och, Mus. MS 554, f.3.

the same for Purcell. On the other hand, there is a pressing need at the moment for informed and up-to-date writing on his music. By and large, the general books on Purcell, from W. H. Cummings's *Henry Purcell 1658–1695* (London, 1881; 2nd edn., 1911) to Margaret Campbell's *Henry Purcell: Glory of his Age* (London, 1993), have been concerned more with the biography than the music. There are, of course, recent studies of particular aspects of Purcell's music, notably Curtis Price's *Henry Purcell and the London Stage* (Cambridge, 1984). But some genres—the odes and the consort music, for instance—have hardly been touched by modern scholarship.

For this reason, I have kept biographical matters in Chapter 1 to a minimum, and I have laid out the book in the traditional way, genre by genre. Chapters 2 and 3 are concerned largely with small-scale domestic music, and in them my emphasis has been on the nature of Purcell's musical language. Chapters 4, 5, and 6 deal with larger-scale public genres, and in them my focus is on structural matters. I was tempted at an early stage to tackle my subject as a series of essays on particular issues of style, in the manner of Silke Leopold's *Monteverdi: Music in Transition* (Oxford, 1991). Rewarding as this would have been, I came to realize that such a treatment depends on a greater understanding of the social and musical context than has yet been achieved in the Purcell literature.

Indeed, my main aim in this book has been to establish a credible context for the various genres to which Purcell contributed. He was an extraordinary backward-looking composer in his youth, whose composition studies apparently ranged across more than a century of English musical history. He was, for instance, the youngest composer to write in a number of genres, such as the secular dialogue, the three-part devotional psalm, the fantasia, the In Nomine, and the pavan. But he tended to imitate old music in limited and specific ways: his In Nomines relate, it seems, to Elizabethan examples, while his three-part fantasias are modelled partly on Orlando Gibbons, and his four-part on Matthew Locke.

Also, the antiquarian strain in Purcell is only one of several. His early music, like that of his contemporaries, is much taken up with French formal patterns, derived mostly from the fashionable dances of the day. This does not mean, of course, that French bottles necessarily contain French wine. Purcell's overtures, for instance, follow the two-section pattern popularized by Lully, but are frequently vehicles for the rich, angular, and dissonant musical language he inherited from the previous generation. For this reason, we should not assume that his apparently 'French style' music should be always be performed with *notes inégales*, overdotting, and the rest of the apparatus of the French performing idiom, though they undoubtedly have their place.

In the 1680s Purcell and his contemporaries gradually adopted a more Italianate idiom. The exact identity of the 'fam'd Italian Masters' referred to in the preface to his 1683 trio sonatas has often been a matter of dispute. But in general it seems that he was more influenced by Italian composers of the previous generation, such as Carissimi, Luigi Rossi, Cesti, Cazzati, and Legrenzi, than his near contemporaries, though this may just have been because it took time for the music of Corelli, Stradella, Bassani, and Alessandro Scarlatti to become known in England.

The Italian influence on Purcell was more far-reaching than the French because it transformed his musical language as well as offering him new formal models. He learnt from them to organize his harmonic thinking in standard sequential patterns and logical modulation schemes. At the same time, his melodic writing became simpler and broader in outline, he began to demand much greater virtuosity from his singers and instrumentalists, he replaced the earlier 'patchwork' structures with extended separate movements, and he began to write for a full Baroque orchestra of brass, woodwind, strings, and continuo. In the process, he diluted the highly coloured and angular idiom of his youth. Purcell was mainly valued for his Italianate music in the eighteenth and nineteenth centuries, and much of his pre-1690 output has remained little known and undervalued up to the present. I have tried to redress the balance to some extent here.

A book of this sort is inevitably heavily indebted to the work of others. In particular, I am grateful to Margaret Laurie, Robert Spencer, Ian Spink, Robert Thompson, Christopher Turner, Andrew Walkling, and Bruce Wood for reading drafts in whole or part, improving it greatly with their detailed criticism. Also, I must thank Andrew Ashbee, Richard Barnes, Clifford Bartlett, Bonnie Blackburn, Michael Burden, Lenore Coral, Lisa Cox, Tim Crawford, Dominic Gwynn, Paul Hopkins, Rosamond McGuiness, Bruce Phillips, Curtis Price, Lucy Roe, Crispian Steele-Perkins, and Neal Zaslaw for helping me in various ways. Once again, my daughter Louise kindly helped me with the index. Some of the material in Chapters 3 and 6 has been drawn from my chapter on consort music in *The Purcell Companion* (London, 1994), edited by Michael Burden, and I am grateful to him and to Faber and Faber for permission to use it here.

P.H.

Colchester
1 May 1994

CONTENTS

LIST OF MUSIC EXAMPLES

(by Henry Purcell unless otherwise stated)

NOTE TO THE READER

ORIGINAL written sources have been transcribed without changing spelling, capitalization, or punctuation, though I have not retained the distinctions between italic, black letter, and roman type in contemporary printed documents. Also, 'y' used as a thorn (as in 'y^{e}' and 'y^{t}') has been rendered as 'th', and the interchangeable letters 'i', 'j', 'u', and 'v' have been modernized. Contractions have been expanded within parentheses; editorial additions are within square brackets. Readers should be alert to the fact that some of the quotations taken from secondary sources might have been modernized more radically. Pitches are indicated using the system in which the open strings of the bass violin are *B♭′ F c g*, and the violin *g d′ a′ e″*. The texts of the music examples are mostly derived from *The Works of Henry Purcell*, published by the Purcell Society and Novello & Co. Ltd; some corrections have been made to unrevised volumes (*Works I*). Unless otherwise stated, all early printed works referred to in the text were published in London, and dates of the first performances of plays are taken from W. Van Lennep (ed.), *The London Stage 1660–1800*, i: *1660–1700* (Carbondale, Ill., 1965).

In Purcell's lifetime the year was normally reckoned from Lady Day (25 March) rather than 1 January. Thus '1689' covers what we think of as 25 March 1689 to 24 March 1690, and dates between 1 January and 24 March in quotations from documents have therefore been rendered using the form '1689/90'. My own references to dates, however, have been modernized. The 'Old Style' or Julian calendar was still used in England until 1752, which means that after 1582, when Pope Gregory XIII instituted the 'New Style' or Gregorian calendar, dates in English documents were ten days (eleven after 28 February 1700) behind those on the Continent. I have retained the old English system of currency: there were twelve pence (*d.*) to the shilling (*s.*), and twenty shillings to the pound (£).

ABBREVIATIONS

(based on those used in *Grove 6*)

AcM	*Acta musicologica*
AnnM	*Annales musicologiques*
CEKM	Corpus of Early Keyboard Music
CSPD	Calendar of State Papers, Domestic Series
DTÖ	Denkmäler der Tonkunst in Österreich
EKM	Early Keyboard Music
EM	*Early Music*
FAM	*Fontes artis musicae*
Grove 6	*New Grove Dictionary of Music and Musicians*
GSJ	*Galpin Society Journal*
INV	a portion of a manuscript written from the end with the volume inverted
JAMIS	*Journal of the American Musical Instrument Society*
JRMA	*Journal of the Royal Musical Association*
JVGSA	*Journal of the Viola da Gamba Society of America*
LSJ	*Lute Society Journal*
MA	*Musical Antiquary*
MB	Musica Britannica
ML	*Music and Letters*
MLE	Music for London Entertainment
MMR	*Monthly Musical Record*
MQ	*Musical Quarterly*
MR	*Music Review*
MT	*Musical Times*
P(R)MA	*Proceedings of the (Royal) Musical Association*
RECM	*Records of English Court Music*
RMARC	*Royal Musical Association Research Chronicle*
PRMBE	Recent Researches in the Music of the Baroque Era
SIMG	*Sammelbände der Internationalen Musik-Gesellschaft*
Works I	*The Works of Henry Purcell*, unrevised volume
Works II	*The Works of Henry Purcell*, revised volume
Z	entry in F. B. Zimmerman, *Henry Purcell 1659–1695: An Analytical Catalogue of his Music* (London, 1963)

Library Sigla

(following the RISM system as used in *Grove 6*)

Great Britain

GB-Cfm	Cambridge, Fitzwilliam Museum
GB-Lam	London, Royal Academy of Music
GB-Lbl	London, British Library, Reference Division
GB-Lgc	London, Gresham College (Guildhall Library)
GB-Lpro	London, Public Record Office
GB-Mr	Manchester, John Rylands University Library
GB-Ob	Oxford, Bodleian Library
GB-Och	Oxford, Christ Church
GB-Ooc	Oxford, Oriel College
GB-Y	York, Minster

Ireland

EIRE-Dm	Dublin, Marsh's Library

Japan

J-Tn	Tokyo, Nanki Music Library, Ohki private collection

United States of America

US-Aus	Austin, University of Texas
US-NH	New Haven, Yale University, School of Music Library
US-NYp	New York, Public Library at Lincoln Center, Library and Museum of the Performing Arts
US-SM	San Mari no, Calif., Henry E. Huntingdon Library and Art Gallery

I
PURCELL'S MUSICAL WORLD

HENRY PURCELL was still a baby when, in the spring of 1660, the English republic collapsed, and Charles II returned to England in triumph. England's musical life had been severely disrupted in the Interregnum. The court, the nation's largest musical institution, had broken up in confusion in 1642 at the beginning of the Civil War, leaving its musicians to fend for themselves. A few, like William Lawes and Henry Cooke, joined the Royalist army. Some went abroad. But most turned to teaching, or just slid into obscurity and retirement. A list in John Playford's *A Musicall Banquet* (1651) of 'excellent and able Masters' available in London to take pupils includes many past and future court musicians, including Henry Lawes, Charles Coleman, and Henry Cooke under the heading 'For the Voyce or Viole', and Christopher Gibbons, John Hingeston, and Benjamin Rogers under the heading 'For the Organ or Virginall'.[1]

At the Restoration the royal music was re-established exactly as it had stood at the outbreak of the Civil War in 1642, with the few surviving members restored to their former posts, and empty places filled with newcomers. They still worked in a number of separate ensembles, each with a discrete role in the daily round of court ceremony. The Chapel Royal was the oldest of these groups; it went back at least to the twelfth century.[2] It provided the king with daily choral services in the chapel at Whitehall, or wherever the court happened to be. Its establishment during the Restoration period is set out in successive editions of Edward Chamberlayne's *Angliae notitiae*, first published in 1669.[3] At its head was the Dean of the Chapel, 'who is usually some grave Learned Prelate, chosen by the King', and a sub-Dean 'or Praecentor Capella[e]', who was often a prominent clergyman with musical interests, such as the mathematician, author, and composer William Holder, sub-Dean

[1] P. A. Scholes, *The Puritans and Music in England and New England* (Oxford, 1934; repr. 1969), 166.

[2] D. Baldwin, *The Chapel Royal Ancient and Modern* (London, 1990).

[3] A. Ashbee (ed.), *RECM* v: *1625–1714* (Aldershot, 1991), 280.

1674–89. Under them were twelve boys and thirty-two gentlemen (consisting of twelve priests and twenty singing men). Among the latter, 'commonly called Clerks of the Chappel', were the Master of the Children, three organists, and the Clerk of the Cheque (effectively the Chapel's secretary); in 1661 they were respectively Henry Cooke, William Child, Christopher Gibbons, Edward Lowe, and Henry Lawes.[4]

Among the secular musicians a distinction was still maintained between the few, who had access to the private apartments of the royal family, taught them music, and played chamber music in varying combinations, and the many, who worked in the public areas of Whitehall Palace. The former consisted mainly of lutenists and/or singers (virtually all singers at the time played the lute, and most lutenists sang), and had been called the Lutes and Voices in Charles I's reign; after the Restoration it was usually known as the Private Music. But there were also a number of viol-players, two violinists (three after Thomas Baltzar was given a post in the summer of 1661), a harper, and several keyboard-players.[5]

The many, the violin band and the various groups of loud wind instruments, were initially re-established as they stood in 1642. But the violin band was soon enlarged from about fourteen to twenty-four in imitation of the French court orchestra, the *Vingt-quatre Violons*. The Twenty-four Violins continued to provide dance music day to day at Whitehall, but was also given several new roles. Groups from it were soon playing in the Chapel Royal, in the king's private apartments, and in the odes that were a feature of ceremonies at the New Year, on royal birthdays, or even on such relatively trivial occasions as the king's return from Windsor after his summer holidays. In the process, the Twenty-four Violins rendered other court ensembles obsolete, such as those members of the Private Music who had played contrapuntal consort music, or those 'Saickbuts and Cornets' who, according to *Angliae notitiae*, had been used 'to make the Chappel Musick more full and compleat.'

In both cases the king seems to have been personally responsible for the change. According to Roger North, he only liked music he could beat time to: he 'could not bear any musick to

[4] E. F. Rimbault (ed.), *The Old Cheque-Book or Book of Remembrance of the Chapel Royal from 1561 to 1744* (London, 1872; repr. 1966), 128.

[5] Ashbee, *RECM* i: *1660–1685* (Snodland, 1986), 27, 219–20; v. 32, 108.

which he could not keep the time, and that he constantly did to all that was presented to him, and for the most part heard it standing'; in particular, he 'had an utter detestation of Fancys,' and 'could not forbear whetting his witt upon the subject of the Fancy-musick'.[6] To be fair, Charles was as musical as most kings—he sang and played the guitar—and was the first English monarch since Henry VIII to have experienced the culture of Continental courts at first hand; he and the courtiers who had shared his years of exile knew that much of the music his court musicians tried to offer him was hopelessly old-fashioned in European terms. As a result, by the end of his reign the Private Music and the royal wind groups had effectively ceased to exist as discrete organizations, and were being used as quarries for extra places in the Twenty-four Violins. Only those groups of instrumentalists who were more functionaries than musicians, such as the trumpeters, the drummers, and the fife-players, retained an independent existence.

The Purcells were not, as far as is known, a family with long-established court connections. Henry and Thomas (probably the composer's father and uncle respectively[7]) are not recorded as members of the royal music until the Restoration, though there is a Buckinghamshire tradition that they were kidnapped to serve as choristers in the Chapel Royal in Charles I's reign.[8] The Master of the Children still had the power to 'press' choirboys for the Chapel Royal, and the Purcell family certainly seems to have had Buckinghamshire connections: Elizabeth Purcell, Thomas's daughter, is buried in Wing parish church, and it has been suggested that the brothers are to be identified with the Henry and Thomas Purcell of Thornborough born in 1627 and 1629 or 1630 to John Purcell, who worked as a carpenter for the Verney family of Claydon House.[9]

Be that as it may, Henry and Thomas Purcell were certainly both members of the Chapel Royal by the time it took part in Charles II's coronation service in Westminster Abbey on 23 April 1661, and were given posts as singer/lutenists in the Private

[6] J. Wilson (ed.), *Roger North on Music* (London, 1959), 350.

[7] Purcell's parentage has long been a subject of debate. In particular, see J. A. Westrup, *Purcell* (4th edn., London, 1980), 305–9; F. B. Zimmerman, *Henry Purcell, 1659–1695: His Life and Times* (2nd edn., Philadelphia, 1983), 331–47; and the review by R. P. Thompson of M. Campbell, *Henry Purcell: Glory of his Age* (London, 1993), in *Chelys*, 22 (1993), 49–50.

[8] Zimmerman, *Purcell: Life and Times*, 336.

[9] Ibid. 335–7.

Music in November 1662, succeeding Angelo Notari and Henry Lawes; Henry Purcell had assisted the aged Italian musician since the summer of 1660.[10] Henry was also installed as a singing man and Master of the Choristers at Westminster Abbey in February 1661, but died in August 1664, before he could advance his career further. Thomas Purcell, however, eventually received five more court posts, making no fewer than seven in all: he was a Groom of the Robes by January 1661; he is listed as 'underhousekeeper' at Somerset House in August 1674; he received a place as one of the composers for the Twenty-four Violins in January 1672 (shared with Pelham Humfrey) and a second one in August 1672, succeeding George Hudson; he also received a second place in the Private Music in December 1674, succeeding John Wilson.[11] It is hard to know whether to interpret this extraordinary case of pluralism as evidence of his outstanding talent, his still at obtaining posts, or just his luck. He seems to have done little to justify his posts as a composer of violin music, for the only pieces known by him are a few Anglican chants and a catch.[12]

We do not know when Henry Purcell junior entered royal service as a choirboy in the Chapel Royal, but it was probably in 1668 or 1669, when he was 9 or 10. The only record of his service there comes in the form of orders dated 17 December 1673 to provide 'Henry Purcell, late child of his Majesty's Chapel Royal, whose voice is changed and who is gone from the Chapel' with two sets of clothes and £30 a year.[13] As a chorister Purcell would have been placed under the authority of Henry Cooke, singer, composer, apostle of Italian vocal techniques, and remarkable choir-trainer. Cooke was responsible for feeding and clothing the children, and with providing them with a basic education, which included teaching them 'the Lattin Tongue and . . . learning them to write'.[14] Purcell may also have attended Westminster School after his period in the Chapel Royal; he was certainly bequeathed money and a mourning ring by its famous headmaster, Dr Richard Busby.[15]

[10] Rimbault, *Old Cheque-Book*, 128; Ashbee, *RECM* i. 38–41; v. 40, 122.

[11] Zimmerman, *Purcell: Life and Times*, 321–6; Ashbee, *RECM* i. 111, 128, 132, 134, 141, 144, 165; v. 63, 69, 71, 139.

[12] A chant is printed in C. Burney, *A General History of Music* (London, 1776–89), ed. F. Mercer (London, 1935; repr. 1957), ii. 381; the catch, 'You that love to drink', is in the Tabley Song Book (see Ch. 2), fo. 47^v.

[13] Ashbee, *RECM* i. 131; v. 162; Zimmerman, *Purcell: Life and Times*, 291.

[14] Ashbee, *RECM* i. 44.

[15] Zimmerman, *Purcell: Life and Times*, 49, 283, 293.

The musical tuition in the Chapel included lessons on the violin, the bass viol, the lute, the theorbo, the virginals, and the organ, as well as, presumably, singing and theory.[16] In later life Purcell was a keyboard-player first and foremost, but he would also have been required to sing in the Chapel Royal when not playing the organ there (see below). He is often said to have been a countertenor, on the evidence of ''Tis Nature's voice' Z328/4 (see Ch. 5), but he is listed among the basses in the list of those singing in James II's coronation.[17] He is also pictured with a viola in two supposedly authentic paintings, now lost, and it would be surprising if he did not play the lute or theorbo as well, for they were the main accompaniment instruments for solo vocal music until the 1690s.[18]

It is often said that Purcell was taught by Cooke himself, or by Pelham Humfrey, but in fact two contemporary Oxford writers state that he was a pupil of Christopher Gibbons: according to Anthony à Wood he was 'bred up under Dr Chr(istopher) Gibbons I think', while the Revd Thomas Ford of Christ Church wrote that he was 'Scholar to Dr Blow & to Dr Xtop[h]er Gibbons'.[19] Blow's epitaph in Westminster Abbey states that he was 'Master to the famous Mr H. Purcell', and Henry Hall, Purcell's friend and fellow Chapel Royal choirboy, printed a poem in Blow's *Amphion Anglicus* (1700), p. ii with the line: 'And Britain's Orpheus learn'd his Art from You.'

Christopher Gibbons, like his father Orlando, was an outstanding keyboard virtuoso. John Evelyn called him 'that famous Musitian', and wrote of him 'giving us a tast of his skill & Talent' on the 'double Organ' at Magdalen College, Oxford in July 1654.[20] But he was less distinguished as a composer. Roger North thought him 'a great master in the ecclesiasticall stile, and also in consort musick', but characterized the style of his fantasias accurately as 've[ry] bold, solid, and strong, but desultory and not without a litle of the barbaresque'.[21] Purcell would doubtless have

[16] e.g. Ashbee, *RECM* i. 39, 56, 57, 77, 116, 128.

[17] Zimmerman, *Purcell: Life and Times*, 123.

[18] W. A. Shaw, 'Three Unpublished Portraits of Henry Purcell', *MT* 61 (1920), 588–90; Zimmerman, *Purcell: Life and Times*, 350, 356–7.

[19] J. D. Shute, 'Anthony à Wood and his Manuscript Wood D 19 (4) at the Bodleian Library, Oxford' (Ph.D. diss., International Institute of Advanced Studies, Clayton, Mo., 1979), i. 241; GB-Ob, MS Mus. E. 17, fo. 40^r^.

[20] *The Diary of John Evelyn*, ed. E. S. de Beer (London, 1955), iii. 109.

[21] Wilson, *Roger North*, 299.

learnt much from Gibbons the organist—assuming, that is, he actually received his lessons. Anthony à Wood wrote that Gibbons was 'a grand debauchee' who 'would sleep at Morning Prayer when he was to play on the organ'.[22] There is direct evidence of his habits in GB-Och, Mus. MS 1142A, fo. 7^v: an autograph organ voluntary has the inscription 'drunke from the Cather(i)ne Wheale. | Christ(opher) Gibbons'.[23]

John Blow was clearly the major influence on the young Purcell's development as a composer, whether or not he gave him formal composition lessons. He was much closer in age to his pupil than Gibbons (born in 1649 rather than 1615), and was much the more wide-ranging and important composer, though it is not clear how much music he had written by the early 1670s, when he is likely to have given Purcell lessons. Blow and Purcell certainly had a close personal and professional relationship; Bruce Wood and Andrew Pinnock have argued that they were 'constantly borrowing from each other, right through from the late 1670s to the end of Purcell's life'.[24]

Matthew Locke is also sometimes said to have been Purcell's teacher, and a letter from the former to the latter is often cited as evidence of their personal relationship.[25] Purcell was certainly influenced by Locke, particularly in his early consort music, and wrote a heartfelt elegy, 'What hope for us remains now he is gone?' Z472, after his death in 1677. However, the letter is probably a forgery: it supposedly belonged to Edward Rimbault, and it does not seem to have been seen by anyone else. Rimbault is suspected of having forged or invented other documents; another case is the manuscript of the 'Address of the Children of the Chapel Royal to the King' ZD120, supposedly written by Purcell in 1670. Locke was a Catholic, and was not therefore a member of the Chapel Royal, though he wrote music for it. So Purcell's contacts with him were perhaps the most frequent and intense after 1673, when he received a secular post at court. By then the teenage composer had probably already mastered his craft, and would have had no further need for formal composition lessons.

22 Shute, 'Anthony à Wood', ii. 109.

23 See C. Gibbons, *Keyboard Compositions*, ed. C. G. Rayner, rev. J. Caldwell (CEKM, 18; 2nd edn., Stuttgart, 1989), pp. xvii, 35–6.

24 A. Pinnock and B. Wood, '"Unscarr'd by Turning Times"?: The Dating of Purcell's *Dido and Aeneas*', *EM* 20 (1992), 381; see also Wood, 'Only Purcell E're Shall Equal Blow', in Price (ed.), *Purcell Studies* (Cambridge, forthcoming).

25 Printed in W. H. Cummings, *Henry Purcell 1658–1695* (2nd edn., London, 1911), 27.

How would Purcell have been taught composition? Some of the treatises available to him and his teachers, such as Thomas Morley's *Plaine and Easie Introduction to Practicall Musicke* (1597), and Charles Butler's *Principles of Musik in Singing and Setting* (1636), teach the traditional skills of adding contrapuntal lines to a given plainsong. Others, such as Thomas Campion's *New Way of Making Fowre Parts in Counter-point* (?1613–14), reprinted in Playford's *Introduction to the Skill of Musick* (1655), and Christopher Simpson's own *Compendium of Practical Musick* (1667), tend to approach the subject vertically rather than horizontally, dealing with intervals and chord formations related to a bass. The same is true of John Coprario's manuscript treatise 'Rules how to compose' (?before *c.*1617), which also exists in a late copy attributed to Blow in Daniel Henstridge's hand in GB-Lbl, Add. MS 30933, fos. 162–71.[26]

The vertical approach was inherent in the basso continuo system, though treatises on continuo playing, such as Locke's *Melothesia* (1673) and Blow's 'Rules for playing of a Through Bass upon Organ & Harpsicon', GB-Lbl, Add. MS 34072, fos. 1–5, deal with extempore harmonization rather than with written composition exercises.[27] This dichotomy between counterpoint and harmony, between the vertical and linear aspects of the music, is, of course, still present in theory teaching today, but was made more difficult at the time by the fact that the two approaches were based on different conceptions of tonality. The former was based on successions of intervals defined according to the medieval system of interlocking hexachords ascending from gamut, while the latter assumed the use of chordal progressions in something akin to the modern system of major and minor keys, though it was rarely properly explained.

No wonder amateurs such as Samuel Pepys and Roger North were confused. Pepys took composition lessons from John Birchensha, but was never able to understand harmony sufficiently to enable him to put a bass to a melody, or to master the figured-bass system; later he invested in the 'musical machine' (preserved in the Pepys Library, Magdalene College, Cambridge) that gave

[26] R. S. Shay, 'Henry Purcell and "Ancient" Music in Restoration England' (Ph.D. diss., University of North Carolina, Chapel Hill, 1991), 13–14; M. F. Bukofzer (ed.), *Giovanni Coperario: Rules How to Compose* (San Marino, Calif., 1952) is a facsimile of Coprario's autograph manuscript.

[27] F. T. Arnold, *The Art of Accompaniment from a Thorough-Bass* (London, 1931; repr. 1965), i. 154–72.

him automatic harmonizations of any permutation of notes.[28] North wrote of children having to learn 'the sour'd and misterious Gamut' by heart 'without the least proffer to them of an explanation of it', and added: 'It were to be wish't that a talent of unfolding secrets in musik, so making them familiar to the understanding of beginners, went always along with the profession of teaching.'[29] Some musicians were still teaching the gamut around 1700. In GB-Och, Mus. MS 350, for example, Richard Goodson the elder (Oxford professor of music 1682–1718) copied simple keyboard pieces and songs for his pupil Catherine Brooks. When he copied the voice-parts of two of Purcell's late songs, 'Who can behold Florella's charms?' Z441, fo. 8^{v}–9^{r}, and 'She that would gain a faithful lover' Z414, 9^{v}–10^{r}, he added the hexachords and their mutations on a separate stave as an aid to solmization.

In fact, amateurs such as Roger North were in some ways better equipped than professional musicians to sort out and expound musical theory, for many of them were members of the Royal Society, or were associated with it, and were therefore versed in what came to be called the Scientific Method. On one level Matthew Locke's furious opposition to Thomas Salmon's reasonable and logical proposals to reform notation (which the Royal Society supported) was the perennial response of the professional to the amateur deemed to be out of his depth.[30] But it also should be seen in the context of the wider conflict at the time between conservative and progressive thought systems, the former characterized by respect for tradition, authority, and rote learning, the latter by an insistence on logical thought, and the rational, dispassionate evaluation of evidence.

Of course, Henry Purcell and his fellow Chapel Royal choirboys had the benefit of individual composition lessons from the leaders of their profession, who would doubtless have taught more by example than by precept. Purcell himself, in the 12th edition of Playford's *Introduction to the Skill of Musick* (1694), wrote that 'the best way to be acquainted with 'em, is to score much, and

[28] R. Luckett, 'Music', in *The Diary of Samuel Pepys*, ed. R. Latham and W. Matthews, x: *Companion* (London, 1983), 278.

[29] Wilson, *Roger North*, 59–60.

[30] L. M. Ruff, 'Thomas Salmon's "Essay to the Advancement of Musick"', *The Consort*, 21 (1964), 266–78; O. Baldwin and T. Wilson, 'Musick Advanced and Vindicated', *MT* 111 (1970), 148–50.

chuse the best Authors'.[31] He was thinking particularly of ground-bass compositions, but his early attempts at writing in a number of genres were probably accompanied by wide-ranging studies of the existing repertory, copied into score where necessary. We have copies in his hand of a fragment of Monteverdi's 'Cruda Armarilli' and an eight-part canon by John Bull on the Miserere plainsong (see Ch. 3), an anthem by Blow and a motet by Cazzati in GB-Lbl, R.M. 20.h.8, and a large collection of pre-Civil War and Restoration church music in GB-Cfm, Mu. MS 88.

Mention of R.M. 20.h.8 and Mu. MS 88 brings us to Purcell's three large autograph score-books.[32] At first sight it looks as if he compiled them as part of a systematic and comprehensive plan to organize and preserve his music in uniform volumes genre by genre. GB-Cfm, Mu. MS 88 has symphony anthems at the front and a sequence of full anthems or verse anthems with organ copied at the back with the book turned round (INV); GB-Lbl, Add. MS 30930 has sacred part-songs at the front and consort music at the back; GB-Lbl, R.M. 20.h.8 has symphony anthems at the front and a sequence of odes and songs at the back.

In fact, appearances are deceptive. Mu. MS 88 differs from the other two books in that Purcell did not start it, and that it contains a sizeable amount of music by other composers. The first seven symphony anthems, by Humfrey and Blow, are in another, unidentified hand, which started the index and dated it 13 September 1677. Purcell evidently began his copying after 10 December of that year, when Blow received his doctorate; he always refers to his teacher as 'Dr' rather than 'Mr'. He added more symphony anthems, by Locke and Blow, and included a number of full anthems by pre-Civil War composers—Thomas Tallis, William Byrd, Orlando Gibbons, William Mundy, Thomas Tomkins, Nathaniel Giles, and Adrian Batten—at the back. Robert Shay has demonstrated that Purcell scored them up from the parts of John Barnard's *First Book of Selected Church Music*, published in 1641.[33] It has been generally assumed that he copied them for study purposes, but the fact that he went to the trouble to correct mistakes, resolve problems in the underlay, and sort out

[31] W. B. Squire, 'Purcell as Theorist', *SIMG* 6 (1904–5), 567.

[32] For Purcell's autographs, see N. Fortune and F. B. Zimmerman, 'Purcell's Autographs', in Holst (ed.), *Purcell: Essays on his Music,* 106–21; Shay, 'Henry Purcell'; R. Thompson, 'Purcell's Great Autographs', in Price (ed.), *Purcell Studies.*

[33] Shay, 'Henry Purcell', 121–9.

inconsistent accidentals suggests that he had performance in mind, or at least the preparation of performing parts.

R.M. 20.h.8, by contrast, consists entirely of music by Purcell, apart from the single pieces by Blow and Cazzati. Purcell seems to have copied the sequences of anthems and secular music more or less in order of composition, starting in 1681; he seems to have ended his contribution to the former in 1685 and the latter in 1689. Like Mu. MS 88, it was probably compiled as part of his court duties; indeed, Robert Thompson suggests that it was issued to him at Whitehall for that purpose.[34]

Add. 30930 consists entirely of music by Purcell, and seems to belong to the personal rather than the public sphere of his activities. It contains domestic music of types that are unlikely to have been performed at court, and Thompson has shown that it started life as a collection of unbound gatherings; it may have remained unbound long after Purcell's lifetime. This means that some of the music could have been copied before the present 'title-page', 'The Works | of Hen: Purcell. | Anno Dom. 1680', was added, and there is evidence that one or two of the consort pieces were added in the late 1680s.

Purcell must also have copied a good deal of music specifically for performance. He was paid £5 in 1675–6 for 'pricking out two bookes of organ parts' for use at Westminster Abbey.[35] They have not survived, but we do have his copy of the Gloria and Sanctus of Blow's Service in G major in a fragmentary Chapel Royal organ-book (see Ch. 4), and his score in GB-Lbl, Add. MS 30932, fos. 52^{r}–55^{v} of Pelham Humfrey's symphony anthem 'By the waters of Babylon' arranged for voices and organ.[36] The latter is one of Purcell's earliest autographs, to judge from the childish handwriting; perhaps he made it while still a chorister for performance in the Chapel. A few years later he copied a number of consort pieces into a set of part-books of which US-NH, Osborn MS 515 is the sole survivor (see Ch. 3). A late autograph of songs, GB-Lgc, MS Safe 3, was probably intended for a pupil (see Ch. 2), as was a recently discovered keyboard manuscript (see Ch. 3). The most recent Purcell autograph to come to light, an anthem by Daniel Roseingrave in score on a single sheet in GB-

[34] Thompson, 'Purcell's Great Autographs'.

[35] Westrup, *Purcell*, 24; see also Zimmerman, *Purcell: Life and Times*, 291–2.

[36] F. B. Zimmerman, *Henry Purcell 1659–1695: An Analytical Catalogue of his Music* (London, 1963), 429.

Och, Mus. MS 1215, may also have had a practical purpose. I have argued that Roseingrave, finding himself without a copy of the work after his move to Winchester in 1681, asked Purcell to transcribe it from a London source, now lost; the composer added enough text to enable a copyist to prepare a set of parts.[37]

It was not unusual at the time for musicians to combine their activities as performers or composers with work as music copyists. The printing of music was still in its infancy in England, thanks mainly to the monopoly on the trade granted by Queen Elizabeth in 1575 to Tallis and Byrd.[38] The monopoly effectively went into abeyance in the 1620s, and only a few collections appeared between then and the early 1650s, when John Playford, established in a shop in the porch of the Temple Church from the late 1640s, began to provide amateurs with a wide range of simple vocal and instrumental music. Before long his customers could buy books of country dances (with their choreographies), continuo songs, catches, Latin motets, psalms, pieces for cittern, gittern, lyra viol, violin, and virginals, as well as sets of simple two- and three-part consort music. But printed music was not yet cheap or plentiful enough for it to be not worth professional musicians forming their own manuscript collections, and the genres used at court, such as the verse anthem and the ode, were mostly too extended and complex to get into print, at least until the market began a period of rapid expansion and development in the 1690s.

The 14-year-old Henry Purcell received his first adult post at court some six months before he was discharged from the Chapel Royal. On 10 June 1673 a warrant was issued to swear him in as 'keeper, mender, maker, repairer and tuner of the regals, organs, virginals, flutes and recorders and all other kind of wind instruments whatsoever'. He was to serve as unpaid assistant to John Hingeston on the understanding that he would obtain the salary that went with the post on Hingeston's 'death or other avoidance', but in the event £30 a year was found for him that December, backdated to Michaelmas; Hingeston did not die until 1683.[39]

It is difficult to know to what extent Hingeston and his deputy were actively involved in making, repairing, and tuning instruments. In earlier times their post had been filled by individuals

[37] P. Holman, 'Henry Purcell and Daniel Roseingrave: A New Autograph', in Price (ed.), *Purcell Studies*.

[38] D. Krummel, *English Music Printing 1553–1700* (London, 1975), esp. 10–33.

[39] Ashbee, *RECM* i. 126, 132.

who were as much craftsmen as musicians, such as Andrew Bassano, a member of the famous family of wind-players and instrument-makers.[40] The wording of some of the itemized bills presented by Hingeston to the Lord Chamberlain's office suggests that he carried out some of the work in person, or at least supervised it. In 1673, for instance, Hingeston charged £2. 10*s*. 'For portage for His Ma(jes)t(ie)s Chamber Organ, & to Bernard Smyth for his charges to Windsor'; the implication seems to be that he himself supervised the moving of the royal chamber organ, while he subcontracted work at Windsor to the organ-builder Bernhard Schmidt (Bernard Smith).[41] Hingeston was certainly directly involved in overhauling an organ in June 1675: he charged £1. 10*s*. 'For my charges to & at Windsor 4 dayes to putt his Maj(es)t(ie)s organ in the Chappell in order ag(ain)st his coming thether'.[42] We do not have detailed bills of this sort submitted by Purcell, but he was certainly capable of tuning church organs; he was paid for tuning the Westminster Abbey organ in 1675, 1677, and 1678.[43] Like many organists then and now, he probably had a good working knowledge of the mechanism of his instrument; in 1686 he was among those asked to judge the newly finished Smith organ in the London parish church of St Katherine Cree.[44]

However, the professions of making and playing instruments were drawing apart during Purcell's lifetime. Before the Civil War it was common for performers to make their own instruments. The Bassano family, who arrived in London from Venice in 1540 to form a recorder consort for Henry VIII, were renowned for their sets of wind instruments and also made lutes and viols, while members of the Lupo family, who worked in the string consort recruited for the English Court in the same year, seem to have made violins.[45] As late as the reign of Charles I a prominent player and composer such as Daniel Farrant not only made stringed instruments, but was also, it seems, involved in inventing and developing new models.[46] Keyboard instruments—complex,

[40] D. Lasocki, 'The Anglo-Venetian Bassano Family as Instrument Makers and Repairers', *GSJ* 38 (1985), 112–32.

[41] Ashbee, *RECM* i. 156.

[42] Ibid. i. 57.

[43] Zimmerman, *Purcell: Life and Times*, 291–2.

[44] D. Dawe, *Organists of the City of London 1666–1850* (Padstow, 1983), 47.

[45] P. Holman, *Four and Twenty Fiddlers: The Violin at the English Court 1540–1690* (Oxford, 1993), 121.

[46] See e.g. id., '"An Addicion of Wyer Stringes beside the Ordenary Stringes": The Origin of the Baryton', in J. Paynter, T. Howell, R. Orton, and P. Seymour (eds.), *Companion to Contemporary Musical Thought* (London and New York, 1992), 1098–1115.

expensive, and time-consuming to make—had always been largely the province of specialist craftsmen. But specialist makers of winds and strings tended to emerge in England only when their instruments ceased to be played exclusively by professionals: viol-makers (John Rose, father and son, and Henry Jaye) appeared towards the end of Elizabeth's reign, violin-makers (Jacob Rayman, Thomas Urquhart, and Edward Pamphilon) about fifty years later, and recorder-makers (Samuel Drumbleby and Peter Bressan) in the Restoration period. The boom in English instrument-making in Purcell's lifetime may have been stimulated by the amateur market, but the ready availability of good locally produced instruments must also have acted as a powerful stimulus to professional music-making.

The next step in Henry Purcell's career came on 10 September 1677, when he was sworn in as a composer for the Twenty-four Violins, one of the posts made vacant by Matthew Locke's death the previous month; for some reason the warrant giving him 20*d.* a day and standard livery of £16. 2*s.* 6*d.* was not prepared until December 1680.[47] Once again, it is difficult to be sure what the job involved. One of his duties would have been, in the words of the 1680 warrant, 'to order & direct them in his course of wayting', which seems to mean that he was in charge of the Twenty-four Violins when he and they were on duty at Whitehall, ready to provide background music at meals or ceremonies, or to accompany dancing. Locke and John Banister, the most important composers associated with the group until then, wrote a good deal of functional dance music in the four-part scorings associated with violin bands, but we do not have much music of this sort by Purcell. The incomplete Suite in G major Z770 and the Chacony in G minor Z730 are just about the only complete examples, though there could be others in a group of fragmentary dances, some copied by the composer, in the bass part-book US-NH, Osborn MS 515.

To judge from his surviving output, Purcell spent most of his time as a court composer writing concerted vocal works rather than dance music. By 1677 members of the Twenty-four Violins were playing regularly in the Chapel Royal, and would also have been required to accompany court odes. Thus we have twenty-seven complete or substantially complete 'symphony anthems'

[47] Ashbee, *RECM* i. 173, 192, 231; v. 79.

(verse anthems with strings) by Purcell, most of which are early works, as well as seventeen court odes, and a number of smaller 'symphony songs' with violins, which may have been written to be performed in the private apartments at Whitehall. The same pattern can be seen in Blow's career. He received a post as a composer to the Twenty-four Violins in August 1682 on the death of Thomas Purcell,[48] but only a handful of consort pieces by him survive, which were probably intended for domestic performance. He seems to have written for members of the Twenty-four Violins almost entirely in concerted vocal music: verse anthems, court odes, and his masque *Venus and Adonis*. After the deaths of Locke and Banister in 1677 and 1679 most of the functional dance and ceremonial music required by the Twenty-four Violins was probably provided by the violinist Nicholas Staggins, who unexpectedly became Master of the Music in 1673–4; his predecessor, the Spaniard Louis Grabu, was apparently debarred from office as a Catholic under the provisions of the Test Act.[49]

Hardly any music by Staggins survives, even though he served as Master of the Music for more than a quarter of a century—he died in 1700, still in office. We know that he wrote a large amount of music in the course of his duties, for there are many payments to him over the years for the expenses of copying his compositions, but it all probably perished with the rest of the accumulated performing material of the royal music in the fire that destroyed most of Whitehall Palace in 1698.[50] By contrast, a large number of the odes and anthems that Blow and Purcell wrote for the court at the same time have survived, and this is because they were copied and preserved by a circle of their colleagues, pupils, and admirers—who evidently did not value Staggins's music.

We have this group, which included William Croft, Thomas Tudway, John Walter, John Church, the Winchester John Reading, and several members of the Isaack family, to thank for the survival of much of the Restoration court repertory.[51] But they copied in score, for they were evidently concerned more with preserving it for posterity than with performing it; we have almost

[48] Ashbee, *RECM* i. 201; v. 81.

[49] Holman, *Four and Twenty Fiddlers*, 298–9.

[50] Ashbee, *RECM* i. 155–6, 191; ii: *1685–1714* (Snodland, 1987), 12, 138, 141; v. 147, 158.

[51] See e.g. P. Holman, 'Bartholomew Isaack and "Mr Isaack" of Eton: A Confusing Tale of Restoration Musicians', *MT* 128 (1987), 381–5.

no contemporary sets of parts for Purcell's major concerted works, which means that we have much less information about the size and disposition of their performing forces than, say, the operas and *grands motets* of contemporary court composers in France. Some original sets of parts survive for Oxford academic odes of the Restoration period, including two by Blow, though it is not clear how much they can tell us about how similar works were performed at court.[52]

It is self-evident that Purcell's birthday and welcome odes were written as part of his court duties, and the same is true of his symphony anthems, a genre that was confined to the Chapel Royal. But what of Purcell's anthems with organ? The complication here is that Purcell was also organist of Westminster Abbey: he succeeded Blow to the post sometime in the accounting year ended Michaelmas 1680.[53] We do not know how much of Purcell's church music was written for Westminster Abbey, but in general the choirs of England's cathedrals and collegiate foundations were in decline at the time, and their repertory was becoming increasingly dependent on that of the Chapel Royal. Westminster Abbey had particularly close links with the Chapel since the two institutions traditionally had a number of musicians in common. Indeed, Purcell himself became one of the three organists of the Chapel in succession to Edward Lowe on 14 July 1682.[54] His duties, set out in a Chapel Royal document of 1663, were as follows:

Of the three Organistes two shall ever attend, one at the organ, the other in his surplice in the quire, to beare a parte in the Psalmodie and service. At solemne times they shall all three attend. The auncientest organist shall serve and play the service on the eve and daye of the solemne feastes, viz: Christmas, Easter, St. George, and Whitsontide. The second organist shall serve the second day, and the third the third day. Other dayes they shall waite according to their monthes.[55]

Change came to the royal music with the death of Charles II in 1685. The traditional view, articulated by Burney, was that the new king presided over a decline in the royal music: 'King James II. was too gloomy and bigoted a prince to have leisure or

[52] Id., 'Original Sets of Parts for Restoration Concerted Music at Oxford', in Burden (ed.), *Performing the Music of Purcell*.

[53] H. W. Shaw, *The Succession of Organists of the Chapel Royal and the Cathedrals of England and Wales from* c. *1538* (Oxford, 1991), 332.

[54] Ashbee, *RECM* v. 80; Rimbault, *Old Cheque-Book*, 17.

[55] Rimbault, 83.

inclination for cultivating or encouraging the liberal arts; nor, indeed, does he seem to have revolved any other idea in his mind, than the romantic and impracticable plan of converting his three kingdoms to the Catholic faith.'[56] This is pure Whig prejudice. There is no sign that the royal patronage of music declined in James's reign, as it certainly did after the Glorious Revolution. In fact, he embarked on a thorough reform of the royal household, the first since the reign of Henry VIII. The archaic system whereby musicians were paid salaries ranging from £20 to £200, fixed by precedent and historical accident rather than an assessment of the worth of the recipient, was swept away, and with it went all the multiple posts and additional fees and pensions that had grown up in the previous 150 years.

With the exception of the trumpeters and the fife- and drum-players, all the secular musicians at court were now members of a single 'Private Music', consisting of the Master of the Music, thirty-three rank-and-file members, and an instrument-keeper. In lists of the group Purcell is described as harpsichordist rather than composer, and is placed with the bass viol-player Charles Coleman the younger under a group of five distinguished solo singers, John Abell, William Turner, Thomas Heywood, John Gostling, and John Bowman.[57] The implication seems to be that Purcell and Coleman acted as the continuo team for 'The vocall part', as the singers were called, possibly in the court odes of the period. At the same time, an attempt was made to reform the chaotic royal finances, which had meant that in Charles II's reign court musicians sometimes had to wait years for their wages. John Hingeston told Pepys in December 1666 that 'many of the Musique are ready to starve, they being five years behindhand for their wages', and that the court harper Lewis Evans had actually died 'for mere want'.[58] In 1685 they were all given a standard rate of £40 a year, and a determined effort was made to pay off the enormous arrears that had accumulated.[59] But by then it was too late. The court had already ceased to be at the centre of England's musical life, largely because its musicians could no longer afford to regard attendance at Whitehall as a full-time job. They had

[56] Burney, *General History*, ii. 379.

[57] Ashbee, ii. 2–3, 122; *CSPD, James II, February–December 1685*, ed. E. K. Timings (London, 1960), 360–1; see also Holman, *Four and Twenty Fiddlers*, 417.

[58] Pepys, *Diary*, vii. 414.

[59] Ashbee, *RECM* ii. 199–218.

been forced to supplement their incomes by teaching, playing in London's theatres, and giving public concerts.

John Banister, the effective leader of the Twenty-four Violins until he was embroiled in a financial scandal in the winter of 1666–7, seems to have started his concerts in the early 1670s. They are first recorded in December 1672, and can be traced in newspaper advertisements until January 1679, only months before his death.[60] In Roger North's writings the impression is given that they were cheap, informal, and unpretentious: they were in 'an obscure room in a publik house in White fryars', and 'most of the shack-performers in towne' came to perform.[61] But a recently discovered printed pamphlet, *Musick; or, A Parley of Instruments, the First Part* (1676), which relates to concerts given by Banister in his 'academy' in Lincoln's Inn Fields during December of that year, shows that he had an astonishing instrumental ensemble, consisting of twenty violins, as well as viols, flutes, recorders, 'pipes', flageolets, 'hoboys', cornetts, sackbuts, lutes, theorbos, harps, guitars, citterns, harpsichords, and organs.[62] There must have been about fifty instrumentalists present, even if the instruments mentioned in the plural in Banister's musical ark were only present in pairs; many of them must have been Banister's court colleagues.

Banister's pamphlet also shows that he was running a school in Lincoln's Inn Fields. An address to the 'Courteous Reader' on the last page lists the 'Arts and Sciences taught and practis'd' in the academy: 'All sorts of Instruments, Singing, and Dancing', as well as French, Italian, mathematics, grammar, writing, arithmetic, painting, drawing, fencing, vaulting, and wrestling. There are several other instances at the time of schools run by musicians or dancing-masters. Jeffrey Banister (probably John's brother, and a fellow member of the Twenty-four Violins) and the Chapel Royal singer James Hart ran a 'New Boarding-School for Young Ladies and Gentlewomen' in Chelsea, where Thomas Duffett's masque *Beauty's Triumph* was performed in 1676 with music by John Banister.[63] A similar enterprise in the 1680s and 1690s in the

[60] M. Tilmouth, 'A Calendar of References to Music in Newspapers Published in London and the Provinces (1660–1719)', *RMARC* 1 (1961), 2–4; see also id., 'Chamber Music in England, 1675–1720' (Ph.D. thesis, Cambridge, 1959), 13–17.

[61] Wilson, *Roger North*, 302–3.

[62] Holman, *Four and Twenty Fiddlers*, 349, 351–2.

[63] W. Van Lennep (ed.), *The London Stage 1660–1800*, i: *1660–1700* (Carbondale, Ill., 1965), 238.

same Chelsea house was run by the dancing-master Josias Priest; he put on Blow's *Venus and Adonis* in 1684, Purcell's *Dido and Aeneas* in 1689, and an unnamed 'opera' in, probably, 1691—which may or may not have been another performance of *Dido*.[64]

As far as we know, Henry Purcell was not directly involved in running enterprises of this sort. But he did play an important role in what was effectively another money-making enterprise for court musicians, the annual St Cecilia celebrations on 22 November. Purcell set the first two datable St Cecilia odes, 'Welcome to all the pleasures' Z339, and 'Laudate Ceciliam' Z329, both in 1683. Court musicians figure frequently among the stewards appointed each year to oversee the proceedings, and they predominate in those cases where the names of individual performers are known. For instance, an early source of the 1687 ode, Giovanni Battista Draghi's setting of Dryden's 'From harmony, from heavenly harmony', gives the cast for, probably, the first performance; of seven solo singers mentioned by name, only one, Anthony Robert, was not a court musician at the time, and he received a post in 1689.[65]

Furthermore, St Cecilia's Day odes were frequently repeated in public concerts, as were some court odes. 'Hail, bright Cecilia' Z328, the 1692 ode, was performed at York Buildings, London's first proper concert hall, on 25 January 1693, while Purcell's 1695 ode on the Duke of Gloucester's birthday, 'Who can from joy refrain?' Z342, was heard at Richmond New Wells on 20 and 27 September 1697.[66] Odes were still being written to commemorate special events in the 1690s, but were being performed more and more in public concerts. For instance, patrons of York Buildings in the winter of 1697–8 could have heard 'a new Pastoral' celebrating the Peace of Ryswick on 29 November, another 'new Pastoral' (?the same piece) on 20 December, and Vaughan Richardson's 'Entertainment of New Music, composed for the Peace' on 16 February.[67] The overtures of odes and semi-operas were often played in public concerts around 1700, to judge from sources that preserve them in concert versions, detached from their parent works.

The decline of the monarchy as a patron of music, and the cor-

[64] R. Luckett, 'A New Source for *Venus and Adonis*', *MT* 130 (1989), 76–80; M. Goldie, 'The Earliest Notice of Purcell's *Dido and Aeneas*', *EM* 20 (1992), 392–400; for the location of the Chelsea schools, see A. M. Laurie, 'Purcell's Stage Works' (Ph.D. thesis, Cambridge, 1961), 16.

[65] Holman, *Four and Twenty Fiddlers*, 426.

[66] Tilmouth, 'Music in Newspapers', 14, 21.

[67] Ibid. 21–2.

responding rise of London's commercial musical life, was hastened by the Glorious Revolution. In 1689 William and Mary initially retained James II's Private Music, though James II's Catholic chapel was abolished, and its singers and instrumentalists went into exile with him or were forced to fend for themselves.[68] There were further reductions the following spring, when the Private Music was reduced to twenty-four musicians and an instrument-keeper.[69] The chief casualties were the five solo singers and their continuo team, including Purcell as harpsichordist. But Purcell still served in the Chapel Royal, retained his place as curator of musical instruments, continued to write court odes, and still, it seems, played the harpsichord at court on an informal basis. Hawkins printed a charming anecdote about an informal concert given by the actress Arabella Hunt and the famous Chapel Royal bass John Gostling, who sang Queen Mary some Purcell songs, accompanied by the composer at the harpsichord.[70] The queen, 'beginning to grow tired', eventually asked Hunt to sing the Scots song 'Cold and raw' to her own lute accompaniment. Purcell, 'seeing her majesty delighted with this tune . . . determined that she should hear it on another occasion', and worked it into the bass of a number of her next birthday ode, 'Love's goddess sure was blind' Z331 of 1692.

The true significance of the 1690 retrenchment seems to be that it marks the moment when musicians began to regard employment at court as just an occasional activity on ceremonial occasions, as their successors were to do in later times. William III was a notorious philistine, whose musical tastes, so far as they can be divined, were confined to a liking for martial instruments. Queen Mary had a true Stuart love of music, but she died in 1694, and the intellectual climate at court and in the Whig government increasingly favoured literature rather than music. Roger North and John Locke, arch Tory and Whig respectively, neatly exemplify the characteristic attitudes of their parties to music, as Mark Goldie has recently pointed out.[71] North was a pupil of Jenkins, a friend of Purcell, an active and accomplished amateur musician, and one of the most informed and perceptive writers on

[68] Ashbee, *RECM* ii. 23–4, 28.

[69] W. A. Shaw (ed.), *Calendar of Treasury Books Preserved in the Public Record Office*, ix: *1689–1692* (London, 1931), 609–10; see also Holman, *Four and Twenty Fiddlers*, 431–2.

[70] J. Hawkins, *A General History of the Science and Practice of Music* (London, 1776; repr. 1853 and 1963), ii. 564.

[71] Goldie, 'Earliest Notice', 398.

music of his time. Locke, by contrast, thought music wasted 'so much of a young man's time, to gain but a moderate skill in it; and engages often in such odd company, that many think it much better spared', and added: 'amongst all those things that ever came into the list of accomplishments, I think I may give it the least place'.[72]

We know nothing for sure about Purcell's political opinions, but we may reasonably conclude that during the reign of Charles II and James II he and his court colleagues were Tories to a man, if only because, like musicians down the centuries, they would have been unwilling to bite the hand that fed them—or promised to feed them. But, like their mythical contemporary the Vicar of Bray, they seem to have had no trouble accommodating themselves to the subsequent change of political and religious direction. Incidentally, much has been made in the past of the possibility that Purcell was a Catholic, or had Catholic sympathies (his father-in-law was certainly one). But if so, it is odd that he set so many anti-Catholic verses as catches. For instance, 'Now England's great council's assembled' Z261 condemns Jesuit plots; 'True Englishmen drink a good health to the mitre' Z284 supports the seven Anglican bishops imprisoned by James II in 1688; and 'Room for th'express' Z270, 'Let us drink to the blades' Z259, and 'The surrender of Limerick' Z278 all celebrate the fall of Limerick in 1691.[73]

1690 was certainly a watershed in Purcell's career. Before, he was essentially a court composer, the author of verse anthems for the Chapel Royal, court odes, and other secular music for the Private Music. After 1690 he led a hectic life in the theatre, and cannot have had time for much else. He wrote his first theatre music in 1680, but for one reason or another his theatrical career did not begin in earnest for another decade, until the production of his semi-opera *Dioclesian* in June 1690. *Dido and Aeneas* was written at some point in the 1680s, but it does not seem to have been performed in public in his lifetime. Little more need be said, except that Purcell was the victim of theatre politics in the last few months of his life: in March 1695 Thomas Betterton and most of the leading actors left the United Company (created by a merger of the two rival companies in 1682) to set up on their own, taking the best singers with them. For some reason Purcell

[72] Goldie, 'Earliest Notice', 398.

[73] P. Hillier (ed.), *The Catch Book* (Oxford, 1987), nos. 83, 107, 96, 81, 100.

remained faithful to the source of the trouble, Christopher Rich, the tyrannical manager of the United Company. But, as Curtis Price has pointed out, it meant that some of his greatest and most mature theatre music was performed by children.[74]

What of Purcell the man? If he sometimes seems a less colourful personality than some of his friends and colleagues it may only be that he was not given to cantankerous polemics, like Matthew Locke, or that, for one reason or another, we do not have the equivalent of Roger North's affectionate memoirs of his teacher John Jenkins, or Samuel Pepys's sharp pen portrait of 'little Pellam Humphrys, lately returned from France': 'an absolute Monsieur, as full of form and confidence and vanity, and disparages everything and everybody's skill but his own'.[75] With Purcell, all we have to go on is the following, from an understandably uncritical poem by 'R.G.' (Richard Goodson the elder?), printed at the beginning of *Orpheus Britannicus*, ii (1702; 2nd edn., 1706; repr. 1721):

> So justly were his Soul and Body join'd,
> You'd think his Form the Product of his Mind.
> A Conqu'ring sweetness in his Vizage dwelt,
> His Eyes wou'd warm, his Wit like Lightning melt,
> But those no more must now be seen, and that no more be felt.
> Pride was the sole aversion of his Eye,
> Himself as Humble as his Art was High.

Few of the anecdotes that circulated about Purcell in the eighteenth century can be relied upon, but one or two have the ring of artistic truth about them, even if they cannot be shown to be literally true. Hawkins's story about Purcell, Queen Mary, and 'Cold and raw' is one, and another was told by the playwright and actor Anthony Aston:

> He [the treble Jemmy Bowen], when practising a Song set by Mr. Purcell, some of the Music told him to grace and run a Division in such a Place. O let him alone, said Mr. Purcell; he will grace it more naturally than you, or I, can teach him.[76]

Beyond this fleeting glimpse of a generous, unassuming personality we have remarkably little to go on. Some have seen Purcell's bawdy catches as evidence of a licentious character, but a glance at virtually any Restoration catch collection will show that they

[74] C. A. Price, *Henry Purcell and the London Stage* (Cambridge, 1984), 16–17.
[75] Pepys, *Diary*, viii. 529.
[76] Westrup, *Purcell*, 76.

are no worse in that respect than the average, though they are well above average as music. Hawkins printed an enigmatic story about Purcell's reaction to the news of the murder of Alessandro Stradella in 1682.[77] On being informed 'jealousy was the motive of it, he lamented his fate exceedingly; and, in regard of his great merit as a musician, said he could have forgiven any injury in that kind'. So far as good, but he goes on to report the comment of an unnamed 'relator': '"those who remember how lovingly Mr. Purcell lived with his wife, or rather what a loving wife she proved to him, may understand without farther explication".' There is no ambiguity in the well-known story, also printed by Hawkins, that Purcell's death on 21 November 1695 was caused by a cold caught while locked out of his house by his wife after a drinking bout.[78] But it is hardly credible, for he lived only a few yards from a number of his Westminster Abbey and court colleagues.

Attempts have even been made to divine Purcell's character from his portraits or his handwriting. According to Franklin Zimmerman, 'we catch a glimpse of the character of a man of genius who has known and tested his powers with full confidence' in the well-known (and rather poor) oval picture at the National Portrait Gallery; 'here, we feel, is a man that it would have been good to know'.[79] All admirers of Purcell's music would agree with that, but it is no more use than, say, the analysis of his handwriting that claimed, among other things, that he was 'a careful, conscientious man, very persistent, endowed with strong power of concentration and strength of will', who 'disciplined and moulded his rich inspirations and imagination, was a master of his craft and, in addition, a very competent but very exacting teacher'.[80] In the end Purcell's character is best revealed in his music, and it is to the music that we now turn.

[77] Hawkins, *General History*, ii. 653–4.
[78] Ibid. ii. 748.
[79] Zimmerman, *Purcell: Life and Times*, 381.
[80] F. H. Walker, 'Purcell's Handwriting', *MMR* 72 (Sept. 1942), 155–7.

II

DOMESTIC VOCAL MUSIC

HENRY PURCELL has always been particularly admired as a song composer. Henry Playford wrote in the preface of *Orpheus Britannicus*, i (1698) that 'The Author's extraordinary Talent in all sorts of Musick is sufficiently known, but he was especially admir'd for the Vocal, having a peculiar Genius to express the energy of English Words, whereby he mov'd the Passions of all his Auditors.'[1] In a poem printed a few pages later Henry Hall, fellow choirboy in the Chapel Royal and organist at Hereford, wrote that his friend 'Each Syllable first weigh'd, or short, or long, | That it might too be Sense, as well as Song.' Playford described *Orpheus Britannicus* on the title-page as 'A | COLLECTION | OF ALL | The Choicest SONGS | FOR | One, Two, and Three Voices, | COMPOS'D | By Mr. Henry Purcell', and the publisher rightly claimed in the second edition (1706; repr. 1721) that it excelled 'any Collection of Vocal Music yet Extant in the English Tongue, and may Vie with the best Italian Compositions'.

Orpheus Britannicus quickly became a classic. A second book 'which renders the First Compleat' was published in 1702; an enlarged edition of the two books appeared in 1706 (repr. 1721), and selections continued to appear throughout the eighteenth century; generations of English musicians got to know their Purcell from the two volumes. But *Orpheus Britannicus* is not representative or comprehensive, despite the claim that it was a collection of all 'the choicest songs'. It consists mainly of late songs, particularly those written for plays after 1690. A few early pieces found their way into the 1698 edition, but most were omitted in 1706, presumably because they were thought to be too old-fashioned; even today much of the pre-1690 domestic vocal music is virtually unknown.

When Purcell wrote his first songs he was contributing to a genre that had developed within the span of a single working life. Three of the most important figures in its early history, Angelo

[1] In the second edition the last phrase reads: 'whereby he mov'd the Passions as well as caus'd Admiration in all his Auditors'.

Notari (?1566–1663), Nicholas Lanier (1588–1666), and Henry Lawes (1596–1662), were still alive at the Restoration, and resumed their places in the royal music—though Notari was apparently in his nineties, and had to be assisted in his duties by Henry Purcell senior. The continuo song was created in James I's reign when the theorbo replaced the lute as the main accompaniment instrument, and the technique associated with it of reading from a figured or unfigured bass line rendered existing types of song obsolete: it was no longer necessary to write out tablature parts for lute songs, or to compose polyphonic viol parts for consort songs. Notari, a Paduan singer-lutenist who came to England in about 1611, was an early exponent of the theorbo; his *Prime musiche nuove* (?1613) is the first English publication with a continuo line.[2]

From the first there were two main types of continuo song. One, common in the lute-song, had the rhythm, character, and sometimes the structure of a dance. During the century the dances chosen as models changed as their choreography went in and out of fashion: from pavan and almand to gavotte and bourée, from galliard to corant, saraband, and minuet. Dance songs are usually settings of light verse, with short lines of regular length, and they frequently achieve a high degree of correlation between poetic and musical accent, line endings and phrase endings, rhyme schemes and matching cadences; they are usually strophic, or would be if the poem as set consisted of more than one verse. Dance songs were always common, but they reached a peak of popularity after the Restoration, when the fashion for songs in minuet rhythm was at its height. 'And for songs', Roger North wrote of Charles II, 'he approved onely the soft vein, such as might be called a step tripla, and that made a fashion among the masters, and for the stage, as may be seen in the printed books of the songs of that time.'[3]

Purcell wrote dance songs throughout his career. Typical examples, all published in Henry Playford's *Banquet of Musick*, ii (1688), are the gavottes 'Sylvia, now your scorn give over' Z420 and 'Phillis, I can ne'er forgive it' Z408, and the minuets 'Ah! how pleasant 'tis to love' Z353 and 'Love's power in my heart shall find no compliance' Z395. The latter, with its swinging tune,

[2] I. Spink, 'Angelo Notari and his "*Prime Musiche Nuove*"', *MMR* 87 (1957), 168–77; see also Holman, *Four and Twenty Fiddlers*, 200–5.

[3] Wilson, *Roger North*, 350.

Ex. 2.1. 'Love's pow'r in my heart' Z395

its simple but telling harmonies, and its delightful trumpet fanfares illustrating the words 'Tararara' and 'Victoria!', shows how effective the genre can be (Ex. 2.1). A number of Purcell's later dance songs are cast in the French *rondeau* form, in which words and music have the pattern ABACA (or AABACAA with the implied repeats). Justifiably famous examples are the minuets 'Fear no danger to ensue' Z626/7 from *Dido and Aeneas* and 'I attempt from love's sickness to fly' Z630/17h from *The Indian Queen*. In all but the shortest the music cadences in the dominant at the double bar (or the relative major in minor-key pieces), and returns in the second half by way of one or more of the related minor keys. This is the familiar modulation plan of late Baroque music, but in the rondeau songs Purcell tended to reverse the pattern, taking in related minor keys in the first episode, and moving to the dominant in the second—so that the return to the first section is marked by a V–I progression.

The other type, the declamatory song, was used for more serious poetry. It was established in James I's reign by Lanier,

Alfonso Ferrabosco II, and Robert Johnson, and achieved its classic phase in the 1630s and 1640s, particularly in the work of Henry Lawes. Declamatory songs were always in duple time, and often had the character of a grave almand or air. But the vocal line mirrored the inflections of speech and illustrated the words with appropriate images, so they are rarely tuneful, though they tend to have more melodic coherence than true recitative, Italian or English. Declamatory songs were in theory through-composed, but in practice strophic examples were not uncommon, particularly in the 1680s, when the declamatory principle became somewhat diluted by a fashion for suave melodic writing in patterns of flowing quavers. By the time Purcell came on the scene the type was more or less obsolete, though there are examples in a group of pieces published as by 'Henry Purcell' in 1678 (see below), and its influence lingered on in the 'recitative' sections of Purcell's extended songs of the 1680s.

Nearly all Restoration songs deal with some aspect of love, usually written from the male point of view, though Purcell occasionally set a 'female' poem ('How I sigh when I think of the charms of my swain' Z374 is an example), and many do not reveal the sex of the poet; a common device was for the lover's grief to be reported at second hand by an observer of unspecified sex. In any case, Purcell does not seem to have been much concerned with matching the standpoint of the poet to appropriate voice types: 'What can we poor females do?' Z518 is set as a duet, for treble and bass voices with continuo. The upper voice-parts of most Restoration songs were printed or written in the treble clef, so that they could be sung equally at the higher or lower octave, though some high-lying parts can only have been sung down an octave by countertenors. In the dialogue 'Jenny, 'gin you can love' Z571/7 from D'Urfey's play *A Fool's Preferment* (1688), for instance, Jenny's part is in the treble clef with the range *g'–c'''* and was presumably sung by a man in drag.

The dialogue was effectively a specialized type of declamatory song, dramatizing a brief exchange between two characters. The genre was also created by Jacobean court composers and was developed by their Caroline successors. At first the characters were usually a nymph or a shepherdess (soprano) and a shepherd (bass), as befitted the pastoral interests of Henrietta Maria and her circle, but before long other subjects were tackled, from the Bible, allegory, or Classical myth. Especially popular were the dialogues between Charon and various characters in legend and history who arrived at the banks of the River Styx to demand his services as ferryman to Hades. They include Orpheus (Robert Johnson, ?Alfonso Balls, and Robert Ramsey), Philomel (William Lawes), Eucomisia—i.e. Lord Hastings (Henry Lawes), Amintor (William Lawes), Hobson (John Hilton), and Oliver Cromwell (Henry Hall).[4]

Purcell's dialogues are among the least-known of all his works, in part because the volume in which they were edited, *Works I*, 22, has long been out of print. Also, they have been overshadowed by his mature theatre dialogues of the 1690s, and they belong to a genre that was obsolete even in Purcell's youth; he must have been the last composer to tackle such well-worn themes as Orpheus and Charon ('Haste, gentle Charon' Z490), Love and Despair ('Hence, fond deceiver' Z492), Horace and Lydia ('While you for me alone had charms' Z524), and Thyrsis and Daphne ('Why, my Daphne, why complaining?' Z525). Most of them are simple and old-fashioned in style. In Strephon and Dorinda ('Has yet your breast no pity learn'd?' Z491), the characters alternate in simple declamatory music, four lines of verse at a time, until they come together in a minuet-like chorus, the bass singing an ornamented version of the continuo. There is not much here that could not have been written by a contemporary of Henry Lawes, and it may be that the piece is much earlier than its first publication, in *The Banquet of Musick*, i (1688).

These three genres, dance songs, declamatory songs, and dialogues, make up most of the song repertory until the 1680s. It was transmitted by manuscripts before the Civil War, but John Playford created a market for printed song-books with a series of handsome folio volumes, starting in 1652 with the anthology *Select Musicall Ayres and Dialogues*. Sequels appeared in 1653, 1659, and, probably, 1663, and Playford also published three books of *Ayres and Dialogues* entirely by Henry Lawes, in 1653, 1655, and 1658; the composite volume *The Treasury of Musick* (1669) consists of the second and third of the former and the third of the latter. Most of Playford's early books also contain a

[4] J. P. Cutts (ed.), *La Musique de scène de la troupe de Shakespeare* (Paris, 1959), nos. 28, 29; I. Spink (ed.), *English Songs 1625–1660* (MB 33; London, 1971), nos. 15, 86; J. Playford, *The Treasury of Music* (London, 1669; repr. 1966), i. 80–1; ii. 109–13; H. Playford, *The Theater of Music*, ed. R. Spencer (MLE A-1; Tunbridge Wells, 1983), ii. 47–51; see also Spink, *English Song from Dowland to Purcell* (2nd edn., London, 1987), 48–53, 291.

section of short tuneful glees in three parts, for two trebles (which could be sung by men an octave lower) and bass with or without continuo. Pieces of this sort were often adaptations of solo songs—John Wilson's *Cheerful Ayres* (Oxford, 1660) were 'First composed for one single Voice and since set for three Voices'—and were evidently the staple diet of convivial music meetings, formal or informal. A manuscript supposedly in John Playford's hand seems to have been used by a music club in Old Jewry in the City of London during the 1650s and 1660s, and items from it were eventually printed by Playford in *Catch That Catch Can; or, The Musical Companion* (1667); a second edition, entitled just *The Musical Companion*, appeared in 1673.[5]

The 1667 *Musical Companion* is of particular interest because it contains the earliest piece ascribed to Henry Purcell, 'Sweet tyranness, I now resign my heart' ZS69 (p. 153). It is a typical glee, simple and unpretentious, scored for two trebles and bass, and cast in the form and style of an almand (Ex. 2.2). The question is: could it really be by Henry Purcell junior rather than by his father (or uncle), as has generally been assumed? At first sight it seems unlikely, for it was credited to an apparently adult 'Mr. Hen. Pursell'. But it was reprinted in a solo version in the 1673 *Musical Companion* (with the catch 'Here's that will challenge all the fair' Z253, which has been accepted as authentic), and again by John Banister and Thomas Lowe in their *New Ayres and Dialogues* (1678), where it is in the company of five other songs ascribed to 'Henry Purcell'. By 1678 Henry Purcell junior had acquired an adult court post, and was rapidly making a name as a composer. On the other hand, there is a copy of one of the pieces published in 1678, 'More love or more disdain I crave' Z397, in the Tabley Song Book, fo. 6^v–7^r, a manuscript bound with the 1653 volumes of *Select Musicall Ayres and Dialogues* and Henry Lawes's *Ayres and Dialogues* at GB-Mr, apparently when they were first published—which seems to mean that this song, at least, must be by Henry Purcell senior.[6] At present, the arguments cannot be taken further, but there is no reason why a talented boy of 8 or 9 could not have composed a simple song such as 'Sweet tyranness', especially given the sort of help the 5- and 6-year-old Mozart received from his father.

[5] I. Spink, 'The Old Jewry "Musick-Society", a Seventeenth-Century Catch Club', *Musicology*, 2 (1967), 25–41.

[6] I am grateful to Robert Spencer for drawing this source to my attention.

Ex. 2.2. 'Sweet tyranness I now resign my heart' ZS69

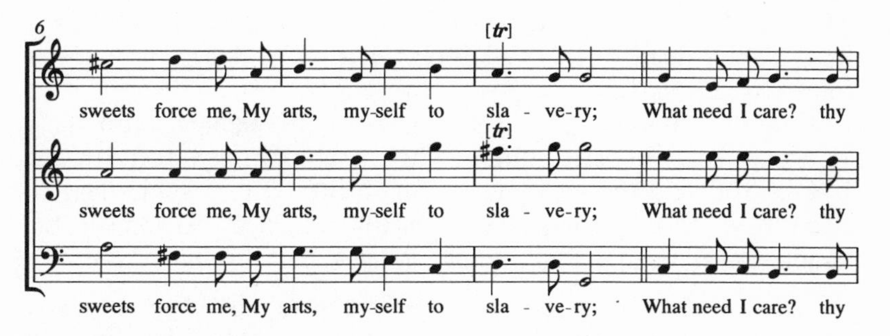

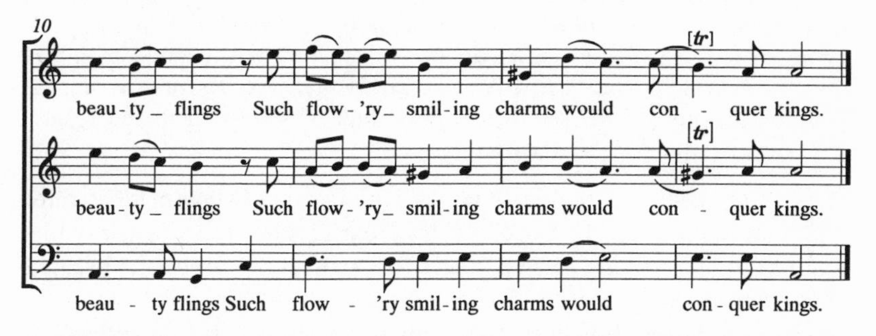

The other genre associated with convivial music meetings is the humble catch. The catch, a special type of round, effectively came into being with John Hilton's *Catch That Catch Can* (1652), and was developed in *The Musical Companion*, the second part of which, also entitled *Catch That Catch Can*, remained in print well into the next century.[7] In catches the phrases are longer than in most rounds, and are separated by cadences, so the first voice completes a defined statement before the second joins in. Also,

[7] For a survey of the genre, see E. F. Hart, 'The Restoration Catch', *ML* 34 (1953), 288–305.

the phrases combined often reveal an unexpected *double entendre*. Most catches deal with wine and women, though a surprising number of other subjects were tackled in the sixty-odd by Purcell, including a wry comment on the ages of man ('An ape, a lion, a fox and an ass' Z241), a loyal toast ('God save our sovereign Charles' Z250), an evocation of Bartholomew Fair ('Here's that will challenge all the fair' Z253), the news ('Is Charleroy's siege come too?' Z257), politics ('Now England's great council's assembled', A catch made in the time of Parliament, 1676 Z261), music ('Of all the instruments that are', A catch for three voices in commendation of the viol Z263), bell-ringing ('Well rung, Tom, boy' ZD107, also attributed to a Mr Miller), and even a publisher's letter to his readers ('To all lovers of music' Z282), printed by John Carr in his *Comes amoris*, i (1687).

Many of Purcell's catches are sexually explicit, and only appeared in a modern edition without bowdlerized texts in the 1970s.[8] An amusing example is 'Sir Walter enjoying his damsel' Z273 (Ex. 2.3). The text is a neatly versified version of the famous anecdote about Sir Walter Raleigh, told by John Aubrey:

> He loved a wench well; and one time getting up one of the maids of honour up against a tree in a wood ('twas his first lady) who seemed at first boarding to be something fearful of her honour, and modest, she cried, 'Sweet Sir Walter, what do you ask me? Will you undo me? Nay sweet Sir Walter! Sweet Sir Walter! Sir Walter!' At last as the danger and the pleasure at the same time grew higher, she cried in the ecstasy 'Swisser Swatter Swisser Swatter.'[9]

It is often assumed that catches and glees were written mostly for amateurs. But many of Purcell's are hard to sing, with fast-moving, angular lines ranging across nearly two octaves, probably because they were actually written for professional musicians off-duty.

By and large, however, the music discussed so far does not require a trained voice in the modern sense. Phrases tend to be short, with few sustained notes; many songs have a compass of only about an octave; and extended passages of florid passage-work are rare. Purcell was heir to a tradition that valued small, sweet voices with clear diction; volume was not admired for its own sake. Charles Butler wrote in 1636 that singers should 'sing

[8] M. Nyman (ed.), *Come Let us Drink* (Great Yarmouth, 1972); also Hillier, *The Catch Book*.

[9] J. Aubrey, *Brief Lives*, ed. R. Barber (2nd edn., Woodbridge, 1982), 265–6.

Ex. 2.3. 'Sir Walter enjoying his damsel' Z273

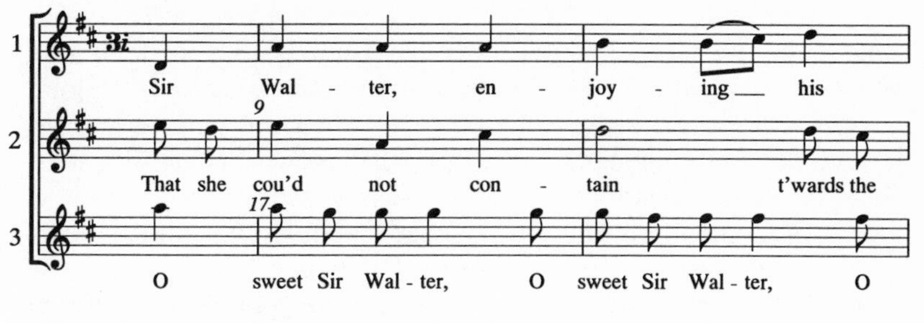

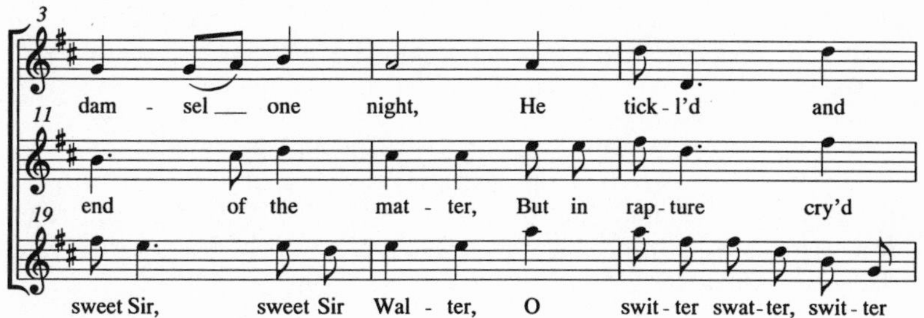

as plainly as they would speak: pronouncing every syllable and letter (specially the vowels) distinctly and treatably'.[10] He asked for 'a sweet melodious treble, or countertenor' soloist in verse anthems, and thought countertenor parts 'fittest for a man of sweet shrill voice'. In John Dowland's translation of the *Micrologus* by Andreas Ornithoparchus (1609), the singer is asked to 'take heed, lest he began too loud braying like an Asse. . . . For God is not pleased with loud cryes, but with lovely sounds.'[11]

[10] E. H. Jones, *The Performance of English Song, 1610–1670* (New York and London, 1989), i. 19.

[11] R. Spencer, 'The Performance Style of the English Lute Ayre *c.* 1600', *The Lute* (1984), 58.

Purcell's contemporaries were of the same opinion. On 8 September 1667 Samuel Pepys heard one Cresset, an amateur, singing in the Chapel Royal 'very handsomely, but so loud that people did laugh at him—as a thing done for ostentacion'.[12] John Evelyn thought the same of Mrs Packer when she sang to Charles II and the Duke of York on 28 January 1685, 'that stupendious Base Gosling, accompanying her'. Her voice was 'so lowd, as tooke away much of the sweetenesse', and Evelyn added: 'certainely never woman had a stronger, or better [voice] could she possibly have govern'd it: She would do rarely in a large Church among the Nunns.'[13] Vibrato, which looms large in modern singing technique, was only acceptable if it was, in Roger North's words:

a gentle and slow wavering, not into a trill, upon the swelling the note; such as trumpetts use, as if the instrument were a little shaken with the wind of its owne sound, but not so as to vary the tone [i.e. pitch], which must be religiously held to its place, like a pillar on its base, without the least loss of the accord.[14]

But it should not be thought that singers of the time were lacking in refinement or virtuosity. Great store was set by the ability to execute ornaments neatly, be they graces (the decoration of a single note according to standard formulae, notated by signs), or divisions (florid ornamentation, extended over several notes). Playford printed a translation of Caccini's instructions for gracing under the title 'A Brief Discourse of, and Directions for Singing after the Italian Manner' in all editions of his *Introduction to the Skill of Musick* between 1664 and 1694, and wrote in the 1666 edition that 'a more smooth and delightful way and manner of singing after this new method by Trills, Grups, and Exclamations' had been 'used to our English Ayres, above this 40 years and Taught here in England, by our late Eminent Professors of Musick, Mr. Nicholas Laneare, Mr. Henry Lawes, Dr. Wilson, and Dr. Colman, and Mr. Walter Porter'.[15] Many pre-Restoration songs survive with added graces and divisions; there are also some ornamented versions of songs by Purcell and his contemporaries, though they tend to use graces rather than divisions.[16]

[12] Pepys, *Diary*, viii. 425. [13] Evelyn, *Diary*, iv. 404. [14] Wilson, *Roger North*, 18.

[15] I. Spink, 'Playford's "Directions for Singing after the Italian Manner"', *MMR* 89 (1959), 130–5.

[16] V. Duckles, 'Florid Embellishment in English Song of the Late 16th and Early 17th Centuries', *AnnM* 5 (1957), 329–45; Jones, *Performance of English Song*, i. 49–144; ii. 1–79.

English vocal music changed rapidly during Purcell's lifetime. A number of Italian singers settled in London in the 1660s and 1670s, bringing up-to-date Italian music with them, and, presumably, a system of voice training that enabled their English pupils to cope with its increased demands. A vocal consort arrived in London around 1663 to establish an Italian opera company. It was led by Vincenzo Albrici (a pupil of Carissimi), and included at least one castrato and the bass singer Pietro Reggio; the harpsichordist Giovanni Battista Draghi joined a little later, and seems to have become a close friend and colleague of Purcell in the 1680s.[17] Later arrivals included Cesare Morelli, Pepys's household musician, and the castratos Giovanni Francesco Grossi (nicknamed 'Siface' after a role in Cavalli's *Scipione affricano*) and Girolamo Pignani. The latter published a book of Italian songs in London, *Scelta di canzonette italiane de più autori* (1679), with pieces by Carissimi, Cesti, Luigi Rossi, Stradella, Pasquini, and others, as well as four Italians on the spot: Bartolomeo Albrici (Vincenzo's brother), Draghi, Nicola Matteis, and Pignani himself.

Reggio's *Art of Singing* (Oxford, 1677) would doubtless have told us more about the way the Italian *émigrés* taught singing, had it survived.[18] In its absence we have to rely on the music they would have taught, such as the pieces in Pignani's collection, and in *Songs Set by Signior Pietro Reggio* (1680). Reggio's setting of 'Awake, awake my lyre' (*Songs*, ii, pp. 5–7), the lyric sung by David to Michal in Abraham Cowley's biblical epic poem *Davideis*, is typical in its demands (Ex. 2.4). It has an angular vocal line including awkward leaps (*e*♭″ to *b*♮′ occurs three times) and several chromatic phrases, and it requires the range of a twelfth, *d*′ to *a*″—a wide tessitura that implies some sort of training to extend the upper range of the voice. Most difficult for the untrained voice, it mixes florid passage-work—needing agility and accurate pitching at speed—with held notes, needing good breath control. Reggio integrates the ornaments into the fabric of the music instead of leaving them to be applied by the singer, so his

[17] W. J. Lawrence, 'Foreign Singers and Musicians at the Court of Charles II', *MQ* 9 (1923), 217–25; J. A. Westrup, 'Foreign Musicians in Stuart England', *MQ* 27 (1941), 72–3, 77–81; G. Rose, 'Pietro Reggio—A Wandering Musician', *ML* 46 (1965), 207–16; M. Mabbett, 'Italian Musicians in Restoration England (1660–90)', *ML* 67 (1986), 237–47.

[18] Spink, 'Playford's "Directions for Singing"', 134; a copy was once in EIRE-Dm; see R. Charteris, *A Catalogue of the Printed Books on Music, Printed Music, and Music Manuscripts in Archbishop Marsh's Library, Dublin* (Clifden, 1982), 132.

Ex. 2.4. Pietro Reggio, 'Awake, awake my lyre', bars 1–17

phrases are much longer than in the average English declamatory song, and require much more stamina.

Italian vocal culture at the time, of course, was largely concerned with the castrato voice. As we have seen, several castrati came to Restoration London, but they remained an exotic novelty: there was no attempt to produce an English race of geldings, nor to assimilate Italian castrati into the mainstream of musical life. But the experience of hearing castrati seems to have inspired

English male singers to develop a distinctive style of alto-range solo singing, which Purcell and his contemporaries exploited. The word 'countertenor' was used at the time to mean parts in this range as well as those who sang them (which is how I use it). Despite many assertions to the contrary, it did not necessarily imply that the singer was a falsetto. Some countertenors undoubtedly were. John Abell, for instance, was described by John Evelyn in January 1682 as 'the famous Trebble'; Evelyn added: 'I never heard a more excellent voice, one would have sworne it had ben a Womans it was so high, & so well & skillfully manag'd.'[19] Most countertenor parts, however, are similar to French *haute-contre* parts (a possible connection that needs to be investigated): they do not go higher than *a'* or *b♭'*, and can easily be sung by light, high tenors, particularly since evidence from woodwind instruments suggests that the secular pitch in England around 1700 was around *a'* = 406.[20] Some countertenors probably blended high tenor and falsetto voices; indeed, it is hard to imagine how else the parts in Blow's Ode on the Death of Mr. Henry Purcell, with the range *d–d''*, could have been sung.[21]

The continuo parts of English songs also began to change rapidly under the influence of the Italians. The bass of Reggio's 'Awake, awake my lyre' is far removed from the simple functional accompaniments that English composers had hitherto used in declamatory songs. It often moves as fast as the voice, imitating it in bursts of semiquavers, and rapidly changing register. Reggio probably played it on the theorbo, but in general the appearance of active bass lines coincides with the move from hand-plucked to keyboard continuo instruments. Singers had traditionally played the lute, and most English keyboard-players did not learn to play continuo until around 1660, to judge from the surviving written-out organ parts in consort music and church music.

But Draghi, Vincenzo and Bartolomeo Albrici, and some of the other Italian singing teachers were keyboard-players, and used harpsichords to accompany secular vocal music. One evening in February 1667, for instance, Samuel Pepys heard the actress

[19] Evelyn, *Diary*, iv. 270.

[20] B. Haynes, 'Johann Sebastian Bach's Pitch Standards: The Woodwind Perspective', *JAMIS* 11 (1985), 78, 100; T. Giannini, 'A Letter from Louis Rousselet, Eighteenth-Century French Oboist at the Royal Opera in England', *Newsletter of the American Musical Instrument Society*, 16/2 (June 1987), 10–11; D. Lasocki, 'The French Hautboy in England, 1673–1730', *EM* 16 (1988), 348.

[21] J. Blow, *Ode on the Death of Mr. Henry Purcell*, ed. W. Bergmann (London, 1962).

Elizabeth Knepp sing 'an Italian song or two very fine' while Draghi 'played the bass upon a Harpsicon'; a few days later he heard the Italian consort sing to two harpsichords.[22] The Playford song-books specify 'Theorbo-Lute, or Bass-Viol' or 'Theorbo, or Bass-Viol' until 1687, when the formula changes to 'Harpsichord, Theorbo, or Bass-Viol' for *The Theater of Music*, iv. By the 1690s harpsichord continuo had become routine, though one should not assume that the bass viol was used as well, as it tends to be today: the continuo group of keyboard and a stringed instrument is characteristic of eighteenth- rather than seventeenth-century England.

Of course, the influence of the Italians was not confined to matters of performance practice. English composers learnt much from their repertory, whether it was imported from Italy or composed on the spot. Its most novel feature was its scale. As part of a general trend away from single-section structures, Italian composers increasingly set long texts as multi-movement pieces, with arias separated by passages of recitative. Nicola Matteis's 'Il dolce contento' (Pignani, pp. 49–55) consists of three sections: a triple-time binary Allegro is followed by a few bars of recitative, which leads to a long duple-time passage marked 'Aria'; after a triple-time Presto there is a return to the opening.

At this embryonic stage in the development of the cantata, movements were usually only contrasted in rhythm: the distinction between recitative and aria verse, the former usually in long unrhymed lines, the latter in short couplets, was only developed by the next generation, as was the characteristic harmonic pattern that placed the arias in a sequence of related keys, linked by rapidly modulating recitative. In fact, in mid-century 'cantatas' tonal contrast is provided more by the arias, with their wide-ranging and increasingly standardized patterns of modulations around a circuit of related keys, than by the recitatives, or by changes of key between movements. As Cesti's aria 'Mia tiranna' (Pignani, pp. 25–31) shows, these patterns also tended to lengthen movements: it runs to more than 100 bars, modulating eleven times in the process. It is a flowing, melodious triple-time piece of the type associated with mid-century Venetian opera (often anachronistically called *bel canto*), and is effectively a da capo aria, with a return to the opening text, music, and key at the end.

An important type of Italian mid-century aria used ground

[22] Pepys, *Diary*, viii. 57, 64–5.

basses. Grounds developed in sixteenth-century Italy as vehicles for dance music, and by extension, for popular vocal music. Strictly speaking, most of them are sets of variations on chord sequences rather than true grounds, for they depend on a standard progression rather than an exact form of the bass line, which often changes from variation to variation or from piece to piece. Furthermore, the Italian repertory of ground basses was disseminated to northern Europe in the early seventeenth century partially in alphabet tablature for guitar, with the chords represented by letters rather than bass lines. The two most common items in this repertory were the *ciaccona* and the *passacaglia*, which became the two most popular grounds in Restoration England. The former, familiar to us from Monteverdi's duet 'Zefiro torna', is a triple-time major-key cadential pattern, while the latter, also in triple time, is a minor-key progression descending by step from tonic to dominant. By Purcell's time the *passacaglia* was often elaborated by making the descending steps chromatic, by adding a cadence, or by turning the ground into the major.

There are no grounds in Pignani's collection, though several ground-bass songs by Italian composers were known in England, and served as models for Purcell and his contemporaries. Reggio's 'She loves and she confesses too' (*Songs*, i, pp. 1–3) is based on the *ciaccona* and was imitated ('satirized' might be a better word; see below) by Purcell, while 'Scocca pur, tutti tuoi strali' is a five-bar *passacaglia* with an added cadence. English sources usually attribute it to 'Baptist', which was taken to mean Giovanni Battista Draghi until Robert Klakowich showed it was by another 'Baptist', Jean-Baptiste Lully; Purcell made a fine arrangement of it for keyboard (see Ch. 3).[23]

Italian vocal music began to make a serious impression on English composers around 1680. The change can be conveniently observed in Playford's second series of song-books, variously called *Choice Songs and Ayres*, *Choice Ayres, Songs, & Dialogues*, and *Choice Ayres & Songs* (*Choice Ayres* for short), which ran from 1673 to 1684. In the first book (1673; 2nd edn., 1675; 3rd edn., 1676) simple strophic dance-songs predominate. The same is true of the second book (1679) and the third (1681), though there are now a few multi-section songs, including the anonymous part-song, 'Mortali che fate' (ii, pp. 63–4), and some extended elegies:

[23] R. Klakowich, '"Scocca pur": Genesis of an English Ground', *JRMA* 116 (1991), 63–77.

William Gregory's 'Did you not hear the hideous groan' on Pelham Humfrey, who died on 14 July 1674 (iii, pp. 49–50), Purcell's 'What hope for us remains now he is gone?' Z472 on Matthew Locke, who died in August 1677 (ii, pp. 66–7), and Blow's 'As on his deathbed gasping Strephon lay' on the Earl of Rochester, who died on 26 July 1680 (iii, pp. 51–2). Ground basses make their first appearance in book iv (1683). John Abell's 'High state and honours' (p. 21) is a *passacaglia* with an added cadence, while Blow's 'Lovely Selina' (pp. 28–9) is a major-key *passacaglia*, and Purcell's 'Let each gallant heart' Z390 (pp. 50–1) a combination of the two types. Purcell's 'She loves and she confesses too' Z413 (pp. 42–3) uses the same *ciaccona* bass as Reggio's setting. 'Lovely Selina' is probably the earliest, for it was written for Lee's play *The Princess of Cleve*, probably first performed in September 1680. The other relevant piece in *Choice Ayres*, iv is Purcell's 'From silent shades' Z370, the mad song Bess of Bedlam (pp. 45–7).

Z472, Purcell's elegy on Locke, is his earliest significant song. It takes the form of an extended declamatory solo followed by a minuet-like 'chorus', in which a second voice sings the bass line. The word 'recitative' is often used in connection with songs of this sort, but it is not very appropriate. In fact, the idiom is essentially that of a declamatory song of the Henry Lawes type, but with a more wide-ranging and more dissonant melodic line, and greatly expanded harmonic horizons. It is characteristic that Purcell uses the modulations in part to illustrate the text according to a well-understood system of key associations.[24] No English musician of the period codified the system on paper, as their contemporaries in Germany and France did, but when Purcell chose D minor for the song he seems to have been responding to a tradition that associated that key with restrained grief. For the same reason, he modulated to F major at the words 'Had charms for the ills that we endure, | And could apply a certain cure', moved suddenly in the direction of G minor at 'From pointed griefs he'd take the pain away', and then to D major for 'his lays to anger and to war could move'. F major, the recorder key (*f′* is the lowest note of the treble recorder), was strongly associated with pastoral peace, just as G minor was associated with grief and death, perhaps because gamut, theoretically the lowest note of the scale,

[24] For a recent discussion of the meaning of keys in Purcell, see Price, *Henry Purcell*, 21–6.

symbolized the grave. Similarly, D major and C major, the two trumpet keys, were associated with ceremony and battle, and B flat, a favourite oboe key, was associated with Bacchic jollity.

To some extent, the system depended on the colours obtained in different keys in unequal temperament, and this is illustrated by 'Urge me no more' Z426. This remarkable song was not printed at the time, perhaps because it was felt to be too experimental; the primary source is in West Sussex Record Office, Cap. VI/1/1, copied by the Eton musician John Walter around 1682.[25] Like Z472 it consists of a declamatory song followed by a minuet passage, and it is also a lament of sorts, though the subject seems to be troubled times rather than the death of an individual; perhaps it refers to the Exclusion Crisis of 1679–82. It is in C minor, a key consistently used by Purcell for tragedy. After initial sallies to E flat major and G minor the music moves steadily round the key cycle in a flatwards direction (B flat minor, E flat minor, A flat major, D flat major, G flat major) to reach a bizarre climax at the words 'horrid outcries of revenged crimes'.

These keys suggest harpsichord continuo: they are more easily encompassed on a keyboard than a fretboard, and the unequal keyboard temperaments of the time, out-of-tune in remote keys, give the 'horrific' words a colour missing in the more equal temperaments used on theorbos.[26] For the same reason, F minor, the most extreme key in regular use at the time, was associated with horror—as in the music for the Sorceress in *Dido and Aeneas*. Purcell may have been encouraged to venture so far round the key cycle by the example of Italian composers, who tended to be more adventurous in these matters; that the journey was accomplished by way of keys a fourth or fifth apart is certainly Italianate, and has little precedent in English music.

Purcell does not use harmony in this song just to provide the necessary forward motion. The words are coloured and illustrated by an angular melodic line, affective harmonies, false relations, and unprepared dissonance. The bizarre final phrase illustrates all these points: the diminished fifth and falling seventh on 'cannot' and 'sobs' (the former making a diminished seventh with the bass), the sudden E♭–E♮ transition on 'my music', and the unprepared fourth with the bass on the word 'speak' (Ex. 2.5). The

[25] H. Purcell, *Secular Songs for Solo Voice*, ed. A. M. Laurie, *Works II*, 25 (London, 1985), p. xix.

[26] A. M. Laurie, 'Purcell's Extended Solo Songs', *MT* 125 (1984), 19–20.

Ex. 2.5. 'Urge me no more' Z426, bars 43–51

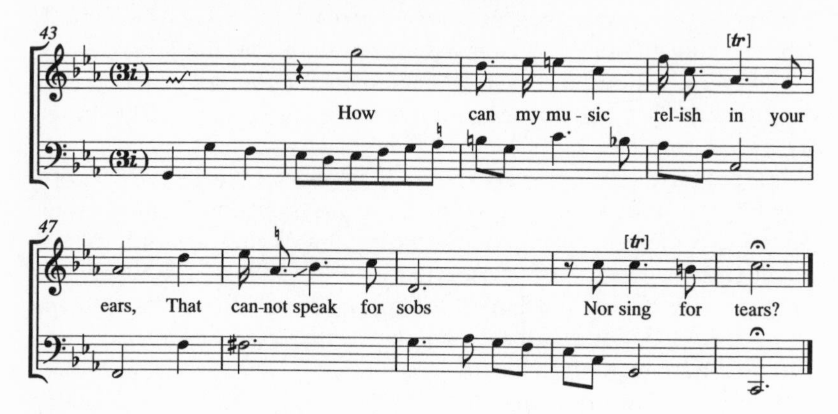

effect is the more piquant when accompanied with the simple rule-of-thumb chords aspiring continuo players were taught; the accompanist should not soften the effect by doubling the dissonant notes.

Purcell did not invent this highly coloured language. It is found, for instance, in the consort music of William Lawes (though, strangely, not in his songs) and Christopher Gibbons, and in much of Locke and Blow. Locke was particularly influential in this respect, and Purcell learnt from him that it was possible to raise the emotional temperature by speeding up a series of conventional progressions—as the last example shows, which touches on five keys in as many bars. But Purcell also used the wider harmonic horizons of Italian music to give his music a sense of direction and space that is lacking in his predecessors.

Purcell's earliest extended multi-section song is Bess of Bedlam, the first of the series of great mad songs that was to end only with 'From rosy bowers' Z578/9, his last song. In Bess of Bedlam the illogical trains of thought of a deranged mind are portrayed by sudden changes of movement and mood, mixing declamatory passages with snatches of airs in duple and triple time. The immediate model is not the cantata, as we might expect, but pre-Civil War antimasque dances that use constant changes of melodic and harmonic direction to portray the antics of comic or bizarre characters. Both words and music of Bess of Bedlam are inspired by the popular ballad Mad Tom of Bedlam, which uses a genuine Jacobean masque tune, 'Gray's Inn Masque'.[27] The voice-part,

'Forth from the dark and dismal cell', was published 'For a Bass alone' in *Choice Ayres*, i, pp. 66–7 (2nd edn., p. 75; 3rd edn., p. 94). A glance at its tune shows that it influenced the shape of parts of Purcell's setting, and explains why he uses an oddly limited and repetitive harmonic scheme, cadencing constantly in C major but sometimes veering unpredictably into C minor. 'Gray's Inn Masque' does the same thing, changing suddenly from G major to G minor, and then back to G major in the middle of a strain. Mad Tom and Mad Bess cast long shadows: their antics inspired a whole genre of mad songs in the plays of the later Restoration period.[28]

As mentioned earlier, 'She loves and she confesses too' is directly connected with Reggio's setting of the same ground bass and the same text, taken from Cowley's *The Mistresse*. Playford wrote in the preface of *Choice Ayres*, iii that he had 'seen lately published a large Volum of English Songs, composed by an Italian Master, who has lived here in England many Years', an unmistakable reference to Reggio. He went on: 'I confess he is a very able Master but not being perfect in the true Idiom of our Language, you will find the Air of his Musick so much after his Country-Mode, that it would sute far better with Italian than English Words.' After accusing Reggio of 'disparaging and under-valuing most of the best English Masters and Professors of Musick', Playford asserted that 'our English Masters in Musick (either for Vocal or Instrumental Musick) are not in Skill and Judgement inferiour to any Foreigners whatsoever, the same Rules in this Science being generally used all over Europe'. Thus Purcell's setting may have been composed as a response and a rebuke to Reggio, probably at Playford's instigation. It certainly contradicts the Italian's word-setting at every turn, and has a more wide-ranging and coherent melodic line, though it too has a few clumsy moments (Ex. 2.6).

In 'She loves and she confesses too' and most of Purcell's early ground-bass songs the bass is a true ostinato, repeated unchanged. Modulation is only possible by cadencing on intermediate steps of the ground, which is one of the reasons why Purcell continually avoids letting the end of a melodic phrase coincide with the

[27] A. Sabol (ed.), *Four Hundred Songs and Dances from the Stuart Masque* (Providence, 1978), no. 152.

[28] See T. Roberts (ed.), *Bring Me Poison, Daggers, Fire: Thirteen Songs of Passion and Madness by Henry Purcell and his Contemporaries* (Oxford, forthcoming).

Ex. 2.6. Pietro Reggio, 'She loves and she confesses too', compared with 'She loves and she confesses too' Z413, bars 1–17

cadence in the bass. In 'O solitude, my sweetest choice' Z406, the finest of the genre, the C minor ground ascends by step from C to A♭ and then curves gracefully into a cadence; it touches on E flat major, F minor, G minor, and A flat major by means of cadences on the third and sixth notes of the bass. In Purcell's later ground-bass songs the music is usually propelled from key to key by a series of modifications to the bass. In 'Now that the sun hath veiled his light' Z193, the serene Evening Hymn that opens Henry Playford's *Harmonia sacra*, i (1688; 2nd edn., 1703; repr. 1714, 1726), pp. 1–3, the G major ground (an ornamented major-key *passacaglia*) is first transposed up a third (taking the music into E minor and B minor), then down a fourth (D major), before returning to the tonic.

Purcell's most advanced ground-bass songs, such as 'Wond'rous machine' in 'Hail, bright Cecilia' Z328/8, have the character of a da capo aria, for the first modulation occurs at the moment when the singer reaches a new phrase of text, and the return to the tonic coincides with a return to the opening words and music. Purcell never wrote any full-blown da capo airs of the sort cultivated by Alessandro Scarlatti and his contemporaries, with their integrated and regularly returning ritornelli and their simple yet brilliant idiom, though a few numbers in his late theatre works use elements of the genre. The incidental music for *The Tempest* once attributed to Purcell includes some Italianate da capo airs, but is now thought to have been written mostly by his pupil John Weldon around 1712.[29]

The next series of song-books, *The Theater of Music* (1685–7), marked John Playford's retirement; Purcell commemorated his death in the autumn of 1686 with one of his most beautiful extended songs, 'Gentle shepherds, you that know' Z464, published separately by 'honest' John's son Henry in 1687. *The Theater of Music* was started by the short-lived partnership of Henry Playford and Robert Carr; they soon fell out, and Playford published the fourth book by himself. But they started with high hopes, and acknowledged the help of Blow and Purcell in one of the prefaces to the first book, which perhaps explains why the tone of the collection is more serious than John Playford's song-books, and why it includes many more extended multi-section songs.

[29] A. M. Laurie, 'Did Purcell Set *The Tempest*?', *PRMA* 90 (1963–4), 43–57.

A number of them are settings of Abraham Cowley (1618–67). Cowley appealed greatly to serious-minded Restoration composers because he retained the lofty tone and extravagant images of the Metaphysicals, but developed a simple, bold, and informal style. The best poetry of the period—Milton, Marvell, and Dryden—tends to be too dense in its meaning and too complex in its phraseology to be suitable for setting to music, and was usually avoided. There had been a Cowley cult among Oxford composers of the 1660s and 1670s, as can be seen from William King's *Poems of Mr. Cowley and Others. Composed into Songs and Ayres* (Oxford, 1668), and Henry Bowman's *Songs for 1, 2, & 3 Voyces* (Oxford, 1677). Reggio included a number of Cowley settings in his *Songs*, and had strong Oxford connections; in the preface he mentions 'Persons very Considerable in that Famous and Flourishing University of OXFORD' among his 'many Worthy Patrons', and Thomas Ford wrote in GB-Ob, MS Mus. E. 17, fo. 41^{r} that he 'livd some time in Oxon after the Restoration'. Cowley became a craze among London musicians following the publication of Blow's 'Awake, awake my lyre' in *Choice Ayres*, iii, pp. 46–8; the work received early performances in Oxford, and may have been written in memory of Christopher Gibbons, who died on 10 October 1676.[30] He was evidently Purcell's favourite poet: we have fifteen Cowley settings among his non-dramatic songs, more than of any other poet; three are in *The Theater of Music*.

'They say you're angry' Z422 (*The Theater of Music*, ii (1685), pp. 20–2), a setting of The Rich Rival from Cowley's *The Mistresse*, is typical of Purcell's early multi-section songs. It consists of three sections: a declamatory opening setting verse 1, a duple-time air with a 𝄵 time signature (implying a fast tempo) setting verses 2 and 3, and a passage of minuet-like triple time setting verse 4. Purcell may well have had Italian cantatas in mind when devising schemes of this sort, but in fact the effect is not very Italianate. The poem does not divide into passages of action and reflection, as cantata texts were beginning to do. It is a continuous outpouring of invective only divided into verses to articulate the rhyme scheme. Thus there is no compelling reason (other than the length of the poem) for Purcell to choose to set one verse

[30] An early set of parts, GB-Ob, MS Mus. Sch. C. 122, partly autograph and partly in the hand of Edward Lowe (Heather professor 1661–82), was evidently copied before Blow received his Lambeth doctorate on 10 Dec. 1677, for Lowe refers to the composer as 'Mr John Blow'; he headed a treble part for the chorus sections, fo. 15^{r-v}, 'for Gibbons'.

in a declamatory style and another as an air, or to divide it into contrasted sections at all, and he makes little of the contrast between his declamatory and tuneful material: the opening has too active a bass to be proper recitative, while the melody of the duple-time air has some declamatory features.

In some of the extended songs of this period, such as 'In a deep vision's intellectual scene' Z545 (?Autumn 1683), a vast setting for SSB and continuo of Cowley's The Complaint, Purcell does use his declamatory and tuneful material in a dramatic way. But this is because The Complaint is a quasi-dramatic poem, a dialogue in reported speech between the poet and his muse. Another example is 'Cupid, the slyest rogue alive' Z367, a setting of a Cowley-like translation of Theocritus. In this small comic masterpiece Cupid receives a taste of his own medicine from a bee in a breathless series of contrasted sections, the erratic changes of direction frequently marking the change from narration to reported speech. In general, however, Purcell preferred the reflective poems of Cowley, particularly those that deal with the joys of the pastoral life, and obtained his contrasts by other means.

Purcell's favourite vehicle for them was the genre popularized by Blow's 'Awake, awake my lyre', a multi-section work for several voices, obbligato instruments, and continuo. The word 'cantata' is often applied to them, but it is peculiarly inappropriate, since they do not depend on the contrast between declamatory and tuneful sections, and they are closer in design and style to the English verse anthem and court ode than to any type of Italian music; they are best called 'symphony songs' by analogy with the 'symphony anthem'. They took up a disproportionate amount of room in printed song-books, and publishers were reluctant to include them. Playford printed the solo sections of 'Awake, awake my lyre' with the apologetic comment 'you have all which is to be sung alone to the Theorbo, and is suitable to the rest in this book', though a few appeared complete in *The Theater of Music* (i–iii of which were advertised on the title-page as containing 'Symphonies and Retornels in 3 Parts . . . for the Violins and Flutes') and subsequent song-books. Of the eight by Purcell in GB-Lbl, R.M. 20.h.8 INV, only two were published at the time: the 'Serenading Song' 'Soft notes and gently raised accent' Z510 for SB, two recorders, and continuo is in *The Theater of Music*, ii, pp. 13–16, while 'How pleasant is this flowery plain' Z543 for ST, two recorders, and continuo, is in *The Banquet of Musick*, i (1688), pp. 41–7. Significantly, they are the two with obbligato recorders: the instrument enjoyed a sudden vogue among amateurs during the 1680s.

Purcell's symphony songs appear in R.M. 20.h.8 INV in the company of his earlier court odes and a selection of his more substantial songs, including 'The Rich Rival' and 'The Complaint'. He seems to have copied the sequence in order of composition, for all the works that can be dated precisely occur in sequence, from Purcell's second court ode, 'Swifter, Isis, swifter flow' Z336 (August 1681), to Queen Mary's birthday ode 'Arise, my Muse' Z320 (30 April 1690), occur in sequence.[31] Thus four symphony songs, 'How pleasant is this flowery plain', the unfinished 'We reap all the pleasures' Z547 for STB, 2 recorders, and continuo, 'Hark how the wild musicians sing' Z542 for TTB, two violins, and continuo, and 'Hark, Damon, hark' Z541 for SB, 2 violins, 2 recorders, and continuo, come after the ode 'The summer's absence unconcerned we bear' Z337 (21 October 1682), and were presumably written over the next six months—the text of Z542 is concerned with the spring. Similarly, 'See where she sits' Z508 for SB, 2 violins, and continuo comes just before the ode 'From hardy climes' Z325 (28 July 1683), while 'Oh! what a scene does entertain my sight' Z506 for SB, violin, and continuo and 'Soft notes and gently raised accent' come between 'Laudate Ceciliam' Z329 (22 November 1683) and 'From those serene and rapturous joys' Z326 (?25 September 1684). 'If ever I more riches did desire' Z544 for SSTB, 2 violins, and continuo is rather later: it comes between 'Ye tuneful muses' Z344 and 'Sound the trumpet, beat the drum' Z335, odes performed in October 1686 and October 1687.

There are many good things in the earlier symphony songs, especially in the Cowley settings 'How pleasant is this flowery plain' and 'See where she sits', but the best is the latest—if the dating process just discussed is reliable. 'If ever I more riches did desire' is also a setting of Cowley; the text is made up of lines from two of the 'Several Discourses by Way of Essays in Verse and Prose', no. 6, 'Of Greatness', and no. 3, 'Of Obscurity'. As befits a text that deals in quasi-religious terms with moral questions—the futility of pride and ambition, the virtue of a humble,

[31] Fortune and Zimmerman, 'Purcell's Autographs', 112–15.

obscure life—the work is virtually a secular anthem. It has a French overture, the most common opening for symphony anthems, and in three of the five vocal sections the words and music given to a solo voice are taken up by the full ensemble, a procedure associated with verse anthems ever since the sixteenth century. The problem with a text that is essentially an Ode to Dullness is that it is hard to avoid setting it either to dull music, or to music that contradicts the text with unwarranted excitement. It is a measure of Purcell's greatness that he avoids both pitfalls. The work is sober and dignified, and yet the interest is easily maintained, in part because the textures are continually varied—between solo and tutti, and between the scoring of the solo sections: soprano and continuo, bass, two violins and continuo, two sopranos and continuo, and tenor, violin, and continuo. The latter, 'Here let my life with as much silence slide', is one of Purcell's most consistently beautiful ground basses.

It is likely that Purcell's symphony songs were written to be performed by members of the Private Music in the royal apartments at Whitehall. They have much in common with court odes, though they are shorter, written for smaller forces, and are less formal in tone. The theme of some of them—a world-weary longing to escape the hurly-burly of public life—was a fashionable one among Charles II's literary courtiers, and it is striking how many of them have texts that refer to the spring. Perhaps they were first heard at informal court celebrations akin to the annual maying ceremonies of Henry VIII's court. It is significant that Purcell marked an alternative low D in the bass part of 'If ever I more riches did desire' 'Gostling', for the Revd John Gostling was essentially a church musician (see Ch. 4), and seems to have confined his secular musical activities largely to the court.

Oddly, the anthem left rather more of a mark on Purcell's symphony songs than on his domestic sacred music. We tend to think of all settings of sacred words as church music, but there was a strong tradition in the seventeenth century of performing anthems and motets at home, and a large repertory of sacred songs and part-songs developed intended mainly or exclusively for domestic use. Roger North wrote that when he was a child his family normally played instrumental music during the week, but that 'on Sunday night, voices to the organ were a constant practise'.[32]

[32] Wilson, *Roger North*, 10.

Thomas Mace wrote in *Musick's Monument* (1676), p. 235 that at music meetings 'we did Conclude All, with some Vocal Musick, (to the Organ, or (for want of That) to the Theorboe', and he added:

The Best which we did ever Esteem, were Those Things which were most Solemn, and Divine, some of which I will (for their Eminency) Name, viz. Mr. Deering's Gloria Patri, and other of His Latin Songs; (now lately Collected, and Printed, by Mr. Playford, (a very Laudable, and Thank-worthy work) . . .

Richard Deering, who died in 1630, probably wrote his Latin motets for the private Catholic chapel of Queen Henrietta Maria where he was organist, but their popularity was not confined to Catholic circles. They were particularly liked by Oliver Cromwell, who had John Hingeston perform them at Whitehall with two boys, according to Anthony à Wood, 'tho(ugh) he did not allow singing or Organ in Churches'.[33] John Playford finally published them in *Cantica sacra* (1662), and brought out a sequel in 1674 consisting mainly of pieces in the same style by Locke, Benjamin Rogers, Christopher Gibbons, and others, as well as some more ascribed to Deering; Playford wrote in the preface that the latter were 'much of Mr Dering's Way, yet by some believed not to be his, but all that have heard them conclude them Excellently Good'.

The 1674 *Cantica sacra* is a representative selection of the domestic sacred music Purcell would have known in his youth. It begins with Latin motets either for soprano or tenor, bass, and continuo, or two sopranos or tenors and continuo (Playford wrote that they 'may properly be Sung by Men as well as Boyes or Weomen'), and continues with 'English Anthems and Hymnes', some of which are cut-down versions of choral pieces—there is, for instance, a soprano and bass version of Locke's five-part verse anthem 'Lord, let me known mine end' (p. 33), the words taken from Ps. 39 in the 1662 Prayer Book. More typical of the domestic sacred repertory as a whole, however, are metrical psalms, such as 'Let all with sweet accord clap hands' by Benjamin Rogers (p. 27), a setting of Ps. 47 in the version published by George Sandys in 1636, or non-liturgical religious poems, such as William Strode's 'Hymn for Good-Friday', 'See, sinful soul, thy saviour's sufferings', set by Isaac Blackwell (p. 37).

[33] Scholes, *Puritans and Music*, 142.

The classics of this repertory were the three-part psalms in the Sandys version by William and Henry Lawes, published in *Choice Psalmes* (1648).[34] They inspired later collections, such as John Wilson's *Psalterium Carolinum* (1657) and a set of manuscript three-part psalms by Locke, and they were still circulating after the Restoration.[35] Pepys, for instance, sang 'some psalmes of Will. Lawes' at the Earl of Sandwich's house on 7 November 1660, and again at Mr Paget's on 12 April 1664.[36] Thus it is not surprising that a number of Purcell's early sacred part-songs are for three male voices (ATB or TTB) and continuo, and that many of them are settings of metrical psalms, two from Sandys and no fewer than nine from John Patrick's *A Century of Select Psalms* (1679).

Patrick was Preacher at the London Charterhouse from 1671 until his death in 1695, and his collection is said to be 'For the use of the Charter-House', so it is possible that Purcell set them for devotional purposes there. Nigel Fortune has drawn attention to several other links between Purcell and the Charterhouse: the text of his anthem 'Lord, who can tell how oft he offendeth?' Z26 is related to one in use at the Charterhouse, and 'Blessed is the man that feareth the Lord' Z9 is said in the main source, the Gostling score at US-Aus (see Ch. 4), to be a 'Psalm Anthem for the Charterhouse sung upon the Founders day. by Mr Barincloe & Mr Bowman'; significantly, this 'anthem' may have started life as a duet for tenor, bass, and continuo.[37] They were certainly written about the time Patrick published his collection, for they seem to be among the earliest pieces in GB-Lbl, Add. MS 30930, possibly copied before Purcell dated it 1680.[38]

Purcell probably had *Choice Psalms* at the back of his mind when deciding which texts of Patrick to set, and how to lay them out. Like the Lawes brothers, he tended to choose gloomy penitential psalms, which offered rich possibilities for his affective musical language. He also followed them in laying out many sections of his three-part pieces in the 'trio sonata' layout of two high parts in sixths and thirds over a sung bass, though he was more inclined than they to vary the texture with lengthy solos,

[34] See M. Lefkowitz, *William Lawes* (London, 1960), 235–49.

[35] For the Locke, see R. Harding, *A Thematic Catalogue of the Works of Matthew Locke* (Oxford, 1971), 3–6.

[36] Pepys, *Diary*, i. 285; v. 120.

[37] N. Fortune, 'Purcell: The Domestic Sacred Music', in Sternfeld *et al.* (eds.), *Essays on Opera and English Music*, 64–5; Zimmerman, *Purcell: Catalogue*, 9–10.

[38] Thompson, 'Purcell's Great Autographs'.

passages of complex counterpoint, or moments where the bass voice separates from the continuo to make an independent part.

Of course, these traits are found in abundance in Restoration church music, and it seems on the face of it that Purcell was creating a new mixed genre, midway between the devotional psalm and the verse anthem. But it may be that some of the pieces by Locke and Blow that are thought of today as anthems really belong to the devotional tradition, and were also Purcell's models. There are some obvious examples in Blow's early autograph score-book, GB-Och, Mus. MS 14, such as his own 'How doth the city sit solitary' (fo. 3^{r}), Christopher Gibbons's 'Ah, my soul, why so dismayed?' (fo. 31^{r}), and Locke's 'O give thanks' (fo. 133^{v}); the latter, like Purcell's 'Blessed is the man', exists in versions with and without a concluding chorus.[39]

However, the immediate model for one of the best of Purcell's three-part psalm-settings, 'Hear me, O Lord, the great support' Z133, seems to have been a true verse anthem, Pelham Humfrey's fine 'Hear O heav'ns', for ATB verse, four-part choir, and continuo.[40] The most obvious point of contact is a highly unusual type of declamation, with the words divided among the voices in a quasi-dramatic way (Ex. 2.7). But Purcell also drew on Humfrey's striking setting of 'Come now, let us reason together saith the Lord', a type of ensemble declamation (related perhaps to the *falsobordone* passages of Italian motets), for 'Fond men, that would my glory stain'. The two works are similar in other ways, not least their harmonic plans, both largely shackled to C minor, Purcell's favourite key for tragedy and pathos. But Purcell far exceeds his model with his daring harmonic writing, such as the grinding dissonances over the pedals at the opening, virtually bitonal at one point.

At the opening of 'O, I'm sick of life' Z140, another fine C minor work, the neurotic effect is achieved not just by conventional falling chromatic lines, but by their juxtaposition, which continually throws up augmented and diminished intervals. Purcell's instinctive grasp of structure led him to end the piece with a rising chromatic passage that balances the opening, the

[39] For Mus. MS 14, see H. W. Shaw, 'The Autographs of John Blow', *MR* 25 (1964), 88–9; J. P. Wainwright, 'The Musical Patronage of Christopher, First Baron Hatton (1605–1670)' (D. Phil. thesis, Oxford, 1992), ii. 165–7; for the Locke, see Harding, *Thematic Catalogue*, 6–7.

[40] P. Humfrey, *Complete Church Music: I*, ed. P. Dennison (MB 34; London, 1972), no. 7.

Ex. 2.7*a*. Pelham Humfrey, 'Hear O heav'ns', bars 34–43

Ex. 2.7*b*. 'Hear me, O Lord, the great support' Z133, bars 33–9

unsettling sharpwards drift of its accidentals—*b*♮, *a*♮, *e*♮, *f*♯—combined with the bass grinding against the continuo—an apt illustration of the land where 'death, confusion, endless night and horror reign' (Ex. 2.8).

Purcell's devotional works in more than three voices tend to be more anthem-like than the trios, mainly because the increased number of parts offers more scope for contrapuntal writing. But they are still clearly intended for single voices. Most of them are scored for SSAB or SSTB rather than the conventional SATB choir, and they have soloistic vocal writing even in the ensemble passages. For instance, in the opening bars of the remarkable fragment 'Ah! few and full of sorrow' Z130 Purcell manages to combine angular and expressive melodic lines with formal counterpoint—in effect, the declamatory song married to the anthem.

This is equally true of the two Latin 'motets', 'Beati omnes qui timent Dominum' Z131 (SSTB and continuo) and 'Jehova, quam multi sunt hostes' Z135 (SSATB and continuo), though Z135 is more popular today with choirs than one-to-a-part consorts. Their models seem to be the group of Latin pieces by Blow in GB-Och, Mus. MS 14, including the remarkable 'Salvator mundi' (fo. 122^{r}), scored, like Z135, for SSATB and continuo.[41] In GB-Lbl, Add. MS 30930 Z131 and Z135 are followed by the opening bars of 'Domine non est exaltatum cor meum' Z102; it is classified as a canon for two voices and continuo in Zimmerman's catalogue, but is probably a fragment of another four- or five-part piece.

Most of Purcell's later sacred songs were printed in Henry Playford's *Harmonia sacra*, i and ii (1693; 2nd edn. 1714; repr. 1726), the equivalent of the Playford series of secular song-books. Some of them suggest that Purcell was steeped in the devotional song repertory. For instance, John Playford's simple setting of 'Close thine eyes and sleep secure', printed in his *Psalms and Hymns in Solemn Musick* (1671), p. 91, seems to have been the starting-point for Purcell's duet Z184 (i, p. 77). Both use the same corrupt version of 'Upon a quiet conscience', a poem often attributed to Charles I but actually by Francis Quarles, and both use the same plan: a duple-time opening, changing to triple time at 'The music and the mirth of kings', and back to duple time at 'Then close thine eyes'.

[41] J. Blow, *Salvator mundi*, ed. H. W. Shaw (London, 1949).

Ex. 2.8. 'O, I'm sick of life' Z140, bars 54–62

Similarly, 'In guilty night' Z134 (ii, p. 40) uses an anonymous text that Purcell would have come across in the popular setting by Robert Ramsey, organist of Trinity College, Cambridge from 1628 to 1644. It is a dramatized version of the encounter between Saul and the Witch of Endor in the Book of Samuel, who summons the ghost of Samuel to tell the king's fortune. Ramsey's setting survives in six sources, several of which date from after the Restoration; two transmit a version ascribed to Nicholas Lanier with essentially the same voice-parts but a modernized continuo. A third setting, by Benjamin Lamb (organist of Eton 1705–33), seems to have been inspired by the Ramsey rather than the Purcell.[42]

A comparison between the settings of Saul and the Witch of Endor is revealing. Purcell followed Ramsey in allocating the parts of the Witch and Samuel's ghost to a soprano and a bass, though his Saul is a countertenor while Ramsey's is a more conventional tenor. Both belong to the declamatory dialogue tradition, though Purcell expanded its technical and expressive range by adding two Italianate features: there are brilliant bursts of semiquavers illustrating phrases such as 'pow'rful arts', 'raise the ghost', and 'ascending from below', and the emotional temperature is continually raised by repetitions of affecting words, such as 'forbear', 'no, no', 'alas!', 'oh!', and 'tell me'. Purcell's setting is also much the more dramatic. Ramsey gives the opening passage of narration to the tenor, but Purcell gives it to all three singers so that Saul's command 'Woman, arise' is his first solo utterance. Ramsey ends rather lamely with the three singers taking up the words and the music of the ghost's spine-chilling verdict:

Art thou forlorn of God and com'st to me?
What can I tell thee then but misery?
Thy kingdom's gone into[43] thy neighbour's race,
Thine host shall fall by sword before thy face.
Tomorrow (till then, farewell, and breathe)
Thou and thy son[44] shall be with me beneath.

Instead, Purcell picked out the word 'farewell', and wrote a last chorus with the singers in character to the end, the witch and the

[42] R. Ramsey, *English Sacred Music*, ed. E. Thompson (EECM 7; London, 1967), no. 10; see also B. Smallman, 'Endor Revisited: English Biblical Dialogues of the Seventeenth Century', *ML* 46 (1965), 137–45; E. Thompson, 'English Biblical Dialogues', ibid. 289–90; M. Chan, 'Drolls, Drolleries, and Mid-Seventeenth-Century Dramatic Music in England', *RMARC* 15 (1979), 126–9; for Lamb, see Shaw, *Succession of Organists*, 375–6.

[43] 'Unto' in the Ramsey setting.

[44] 'Sons' in the Ramsey setting.

ghost icily dismissing the shattered king, who can only sob in semitone slides over descending *passacaglia*-like harmonies (Ex. 2.9). Once heard, these wonderful yet horrifying ten bars can never be forgotten.

Ex. 2.9. 'In guilty night' Z134, bars 140–9

There are many fine pieces among Purcell's contributions to *Harmonia sacra*. 'Lord, what is man?' Z192 (ii, p. 1) comes to mind, as does the virtuosic setting of Nahum Tate's Hymn upon the Last Day, 'Awake, ye dead' Z182 (ii, p. 71), rarely performed because of its scoring for two basses and continuo. But two are of particular interest because, as it were, they extend the tradition of biblical dialogues to solo song. 'Let the night perish', (Job's Curse) Z191 (i, p. 10), and 'Tell me, some pitying angel', (The Blessed Virgin's Expostulation) Z196 (ii, fo. 4^v), are just about the only English devotional songs that represent the point of view of a single biblical character. 'Job's Curse', a verse paraphrase by Jeremy Taylor of Job 3, is a relatively conventional declamatory song with a triple-time passage at the end portraying the quiet of the grave, though the twists and turns of the minuet-like music suggest that Purcell did not view the prospect entirely with Christian resignation.

The Blessed Virgin's Expostulation is perhaps the closest Purcell came to writing an Italianate cantata. It is a setting of words by Nahum Tate, subtitled in *Harmonia sacra* 'When our Saviour (at Twelve Years of Age) had withdrawn himself, &c. Luke 2. v. 42.', and is an astonishingly vivid and human portrayal of a mother who has lost her child, her words veering rapidly between hope and despair. Tate's libretto for *Dido and Aeneas* has been much mocked, but we should recognize that his extravagant gestures and flowery images were the poetic equivalent of Purcell's increasingly Italianate musical language, and gave the composer the stimulus he needed. Tate evokes a mind in turmoil by means of a startlingly modern series of disconnected statements, commands, and questions, expressed in simple, direct language, and aptly laid out in verse of shifting metre and rhyme.

Purcell seizes the opportunity with abandon, using his full repertory of affective devices (including, most memorably, the cries of 'Gabriel!' set to high repeated Gs over a sequence of increasingly clashing *passacaglia*-like harmonies), and he divides the setting into a sequence of declamatory passages and binary airs—equivalent to Recitative–Aria–Recitative–Aria–Recitative in a cantata. But again, the pattern is implied to some extent by Tate's text. For instance, the first air, 'Me Judah's daughters', is a moment of reflective reminiscence, and Purcell could hardly help setting the next line 'Now (fatal change!) of mothers most distress'd' as a change from air to recitative, and from major to minor. It is interesting, however, that such a forward-looking piece should use a conservative harmonic plan, alternating passages in C minor and C major. Purcell was surprisingly reluctant to abandon these older key relationships, between major and minor versions of the same tonic, in favour of the more modern ones used in Italian cantatas, which exploited the connection between major keys and their relative minors a third lower.

Purcell spent most of his time and energy after 1690 writing music for the theatre, which is why most of his late songs were written for plays. Theatre songs are heavily in the majority in his late autograph volume of songs, GB-Lgc, MS Safe 3, apparently copied between 1692 and 1695 partly with a pupil in mind; he or she was evidently a soprano, for the collection contains a number of arrangements of pieces originally written for lower voices.[45] There are a few extended multi-section pieces among Purcell's late 'single songs'—the seven-section 'Love, thou can'st hear' Z396 is a spectacular example—but most of them are shorter and simpler, consisting either of a single section, or two contrasted sections, with a declamatory passage frequently followed by a dance-like air.

Some of the former might be categorized as dance songs if they were not so extended and sophisticated. 'I love and I must' Z382, datable from its position in the Gresham MS to the autumn of 1693 and enigmatically subtitled there 'Bell Barr' (see Ch. 3), is in minuet rhythm, but it is over 100 bars long, has some florid passage-work and a wide-ranging modulation scheme, and a bass that converses on equal terms with the voice-part for much of the work. Good examples of the latter are 'The fatal hour comes on apace' Z421, a fine song that only found its way into print in *Orpheus Britannicus* (ii, p. 30), and 'Love arms himself in Celia's arms' Z392, a setting of verses by Matthew Prior.

Z392 is a particularly striking example. The structure of the poem, two complementary pairs of four-line stanzas, must have encouraged Purcell to use the double-barrelled pattern. The first pair, where the poet complains of the effect of Celia's eyes on his reason, is balanced by the second, where he asks 'cruel reason' to argue his case in her heart. Appropriately, Purcell offers both contrast and balance in his two sections. Their overall mood is derived from images in the opening lines. The vocal line at the

[45] W. B. Squire, 'An Unknown Autograph of Henry Purcell', *MA* 3 (1911–12), 5–17; Zimmerman and Fortune, 'Purcell's Autographs', 115–18.

opening, setting the words 'Love arms himself in Celia's eyes' in the fashionable C major fanfare idiom of trumpet music, is anticipated by a phrase in the bass, which strides majestically up to *g′* before cadencing—an excellent example of the increased importance of the bass in the vocal music of the period, and a sign, incidentally, that Purcell was thinking of harpsichord continuo, for the line is too high at that point for a realization to be added on the theorbo or arch-lute. The second sets the words 'Then, cruel reason, give me rest' in C minor, and in a more languishing minuet-like triple time. Each section is articulated at the end of the fourth line by a modulation (to the dominant and the relative minor respectively). Also, the fanfares at the opening are matched in the second half by a rising arpeggio figure at the words 'Go try thy force', and trumpet-like coloratura at 'How great, how godlike 'tis to save'. Similarly, a flatwards lunge in the vocal line at the words 'weak reason' matches the sudden introduction of D♭s and C♭s at the word 'cold' in the second half.

Purcell's late songs may be Italianate in their decorative details, but they are as characteristic of English Baroque art as the buildings of Hawksmoor or the plays of Congreve (or, for that matter, the poems of Prior) in the creative balance struck between symmetry and asymmetry, unity and diversity, simple form and complex patterning, elegance and passion. Not surprisingly, pieces of this sort became the models for much of the serious song repertory of the next generation.[46]

[46] H. D. Johnstone, 'English Solo Song, *c*.1710–1760', *PRMA* 95 (1968–9), 74–6.

III

INSTRUMENTAL MUSIC

PURCELL was heir to rich and distinctive traditions of instrumental music that went back to Tudor times. Around 1600 the consort repertory consisted mainly of fantasias in three, four, five, and six parts, intended for various combinations of treble, tenor, and bass viols, though it also included the graver sort of dance music, and motets and madrigals were often played in instrumental versions. Around 1620 a group of court composers under the patronage of Prince Charles began to experiment with new types of consort music, combining the chamber organ and the violin with viols for the first time. The violin had hitherto been associated almost exclusively with dance music, so Orlando Gibbons and Thomas Lupo wrote fantasias for it in a dance-like idiom, and John Coprario devised the fantasia suite, a fixed sequence of fantasia–almand–galliard, scored for one or two violins, bass viol, and organ. William Lawes, John Jenkins, and others added to the fantasia suite repertory in the reign of Charles I, and devised a number of new scorings; by about 1640 they had begun to supersede conventional viol consorts, at least at court.

The Civil War disrupted the settled pattern of musical life, forcing court musicians to fend for themselves. But consort music flourished. 'During the troubles', Roger North wrote,

> when most other good arts languished Musick held up her head, not at Court nor (in the cant of those times) profane Theaters, but in private society, for many chose to fidle at home, than to goe out, and be knockt on the head abroad; and the enterteinement was very much courted and made use of, not onely in country but citty familys, in which many of the Ladys were good consortiers; and in this state was Musick dayly improving more or less till the time of (in all other respects but Musick) the happy Restauration.[1]

The Puritans were not against music as such, as used to be thought, only against its cultivation in cathedrals and collegiate foundations, and in the theatre.[2] Anthony à Wood even thought

[1] Wilson, *Roger North*, 294. [2] Scholes, *Puritans and Music*.

they 'used to encourage instrumental musick', and had a political motive for doing so: 'but vocall musick the heads of these parties did not care for, and the juniors were afraid to entertaine it because [it was] used by the prelaticall party in their devotions'.[3] The upheaval certainly created the conditions for the dissemination of the violin and its court repertory into the wider musical community. By the end of the Commonwealth amateurs such as Anthony à Wood and Samuel Pepys were playing the instrument, and the viol consort was rapidly becoming obsolete.[4]

Consort music declined after the Restoration. Professional musicians went back to their posts, and were once more taken up with writing and performing vocal music, in church, at court, and in the theatre. At Oxford, Wood noticed that 'when the masters of musick were restored to their several places . . . the weekly meetings at Mr Ellis's house began to decay, because they were held up only by scholars, who wanted directors and instructors &c'.[5] At court, the pre-war group that had played fantasia suites was initially revived under a new name, 'The Broken Consort'; Matthew Locke probably wrote his eponymous suites for it in 1661.[6] But Charles II, with his 'utter detestation of Fancys', soon dispensed with its services. Instead, he granted members of the Twenty-four Violins access to the Privy Chamber, so the orchestral dance music they played became the main type of consort music cultivated by court composers, though the older forms lingered on for a while outside the court.[7]

The repertory of the Twenty-four Violins is most easily recognized by its scoring. The group played in four parts, with a single violin, two violas, and bass, until the middle of the 1670s, and thereafter with the more modern 'string quartet' layout, though Purcell and Blow were still inclined to write second violin parts that function more as an alto than a second soprano.[8] The violin parts of domestic consort music, by contrast, are usually equal in range, and frequently cross, sharing the melodic material. The Twenty-four Violins used bass violins, larger than the cello and tuned *B*♭′, *F*, *c*, *g*. Thus orchestral bass parts frequently descend

[3] Shute, 'Anthony à Wood', ii. 105.

[4] On this point, see Holman, *Four and Twenty Fiddlers*, 267–75.

[5] Ibid. ii. 103.

[6] M. Locke, *Chamber Music II*, ed. M. Tilmouth (MB 32; London, 1972), 1–30; Holman, *Four and Twenty Fiddlers*, 275–6.

[7] Ibid. 284–6.

[8] Ibid. 316–18.

to *B*♭′ and do not go above *d*′ or *e*′, the comfortable upper limit of the bass violin, while bass parts in consort music, intended for viol rather than violin, do not go below *C* and sometimes go up to *g*′ or even higher. It is unlikely that bass instruments at 16′ pitch were used in English orchestras in Purcell's lifetime; by and large, the bass lines of his orchestral music are low enough as it is.

Violin bands had traditionally played without continuo instruments, except when they took part in concerted vocal works. In England, they probably often played without continuo throughout Purcell's lifetime, which is why his theatre suites were published without a separate continuo part or any figures in the bass (see Ch. 6). Most Restoration consort music is scored for one, two, or three violins and bass, and needs continuo instruments to fill the yawning gap between the top and the bottom. The sources often lack specific continuo parts, but we must remember that theorbo-players traditionally played from unfigured basses, and organists accompanied from score; indeed, many scores of consort music (including Purcell's autograph, GB-Lbl, Add. MS 30930) were probably copied partly for that purpose.

The domestic consort and orchestral repertories can also be distinguished to some extent by genre. Fantasias and pavans belong to the contrapuntal tradition, while overtures and chaconnes are essentially orchestral forms, though they are often found in cut-down form in domestic manuscripts. Of course, some dances, such as the corant and the saraband, are found in all types of consort music, though they are combined in different ways. Domestic consort suites tend to be relatively short, with an opening fantasia or pavan followed by three or four dances, while suites written for the court activities of the Twenty-four Violins usually consist of a large number of short movements, all in the same key.[9] Members of the Twenty-four Violins were also regularly assigned to London's theatres in the 1660s and 1670s, and the suites of incidental music they played are similar, though they usually do not keep to a single key. They were not heard in the theatre in a continuous sequence, and traditional key associations were sometimes used to reflect the changing moods of the parent play.

The criteria I have just formulated allow us to identify some of Purcell's early consort pieces as orchestral music, written for the Twenty-four Violins. An obvious candidate is the recently discov-

[9] Holman, *Four and Twenty Fiddlers*, 319–26.

ered Staircase Overture in B flat (not in Z). The only complete version is in a score at Tatton Park in the hand of Philip Hayes (1738–97), but the bass is in Purcell's autograph in the part-book US-NH, Osborn MS 515, and it seems that Hayes copied it from a set of which Osborn MS 515 is the sole survivor; it belonged to Charles Burney, and it still consisted of three part-books as late as 1848, when it appeared in the sale of the library of the Revd Samuel Picart.[10] Hayes's score has parts for two violins, bass, and a separate, simplified continuo, and it is clear from the number of chords without thirds that an inner part has been left out, as commonly happened when orchestral music was copied into domestic manuscripts; there is an editorial viola part in the 1990 revision of *Works II*, 31. The continuo part may have been added by Hayes, perhaps to compensate for the missing viola. Its figuring does not seem to be original or in its original state, for it uses natural signs, which only began to be used by English copyists around 1700.

The Staircase Overture is almost certainly Purcell's earliest surviving consort piece. It opens with rushing scales (hence, presumably, its title) that recall similar passages in a number of pieces written by Locke for the Twenty-four Violins, notably in the first movement of his incidental music for *The Tempest* (1674), and in the storm passage of its famous Curtain Tune.[11] The style of the music suggests that it was written about then. Some of the part-writing is rather aimless, particularly in the minuet-like second section, and the first section is similar to an almand in style. This remained the most common pattern of overture in England until about 1680, when Lully's mature examples of the form began to circulate in manuscript, popularizing the familiar dotted rhythms and searching harmonies in the first section, and a fugue in the second.

Another indication that 1674–5 is about right for the Staircase Overture is provided by the last section, which plunges unexpectedly and eloquently into B flat minor (Ex. 3.1). The model for this was probably the prelude to Robert Smith's suite 'New Years Day', US-NYp, Drexel MS 3849, pp. 47–50, which starts with

[10] A. Browning, 'Purcell's "Stairre Case Overture"', *MT* 121 (1980), 768–9; R. Ford, 'Osborn MS 515: A Guardbook of Restoration Instrumental Music', *FAM* 30 (1983), 174–84; H. Purcell, *Fantazias and Miscellaneous Instrumental Music*, ed. T. Dart, rev. M. Tilmouth, A. Browning, and P. Holman, *Works II*, 31 (London, 1990), pp. xiv–xv; *Catalogue of the Musical Library, etc. of the Late Rev. Samuel Picart, 10 March 1848*, GB-Lbl, S.C.P.6 (1), lot 208; it was sold to 'Wilkes' for 2*s*. 6*d*.

[11] M. Locke, *Dramatic Music*, ed. M. Tilmouth (MB 51; London, 1986), 19–20, 27–9.

Ex. 3.1. The Staircase Overture, bars 25–39

similar music in the same outlandish key. It was probably written for court ceremonies on 1 January 1674 or 1675: Smith became a court musician in 1673, but died unexpectedly in the autumn of 1675.[12]

Osborn MS 515 also contains some other unique pieces that can be attributed with more or less certainty to Purcell; Philip Hayes did not score them up, so only the bass part survives.[13] Four are in Purcell's autograph: a pavan in F minor followed by a minuet-like movement in the same key, another pavan in F minor (this has a second strain in triple time, so it must have sounded more like an overture than a true pavan), and an overture in C major. All of them probably date from the same period, for Purcell copied them using his early type of bass clef, shaped like a reversed S; he went over to a more modern form in GB-Cfm, Mu. MS 88 INV, started no earlier than the autumn of 1677.[14] Three more pieces apparently by him—another overture in C major, a short single-section prelude in B minor, and an almand-like piece also in B minor—were added by the manuscript's compiler, who may have been an associate of the bass singer John Gostling, either in London or Canterbury.

Osborn MS 515 is virtually the only manuscript that preserves Purcell's early consort music in part-book format. The rest are scores: the autograph Add. MS 30930 and a handful of secondary sources from Purcell's immediate circle. GB-Lbl, Add. MS 33236, GB-Lbl, R.M. 20.h.9, and US-NYp, Drexel MS 5061 contain pieces apparently copied directly from Add. MS 30930 or from lost autographs, and the copyist of Drexel MS 5061 was probably one of the Isaack family of organists and choirmen, two of whom were choirboys in the Chapel Royal.[15] R.M. 20.h.9 has several unique Purcell items, and was copied by the John Reading who was organist at Winchester (1675–81 in the cathedral, and 1681–92 at the college).[16]

R.M. 20.h.9 INV opens with four pieces by Purcell. Three are overtures, and belong to the orchestral repertory. One is a variant

[12] Holman, *Four and Twenty Fiddlers*, 325–6.

[13] Edited in *Works II*, 31, 106–9.

[14] The two forms can be seen on fo. 141^r INV, reproduced in Holst (ed.), *Purcell: Essays on his Music*, pl. 3.

[15] For Add. MS 33236, see R. P. Thompson, 'English Music Manuscripts and the Fine Paper Trade 1648–1688' (Ph.D. thesis, London, 1988), 444–53; for the Isaacks, see Holman, 'Bartholomew Isaack', 381–5; M. Locke, *The Rare Theatrical, New York Public Library, Drexel MS 3976*, ed. P. Holman (MLE A-4; London, 1989), pp. xi–xii.

[16] P. Holman, 'Henry Purcell and Daniel Roseingrave'.

(and apparently revised) version of the symphony to the court ode 'Swifter, Isis, swifter flow' Z336/1; it appears as an independent piece in *Works II*, 31. The other two, the four-part Overture in D minor Z771 and the five-part Overture in G minor Z772, may also come from lost vocal works. Z771 probably dates from the early 1680s. It is an example of the mature French type, yet its dimensions are relatively modest, and the part-writing still harks back to the violin–two viola scoring of the early Restoration period: the second part hardly ever crosses the top part, and only once goes out of the range of the viola in first position. Z772 has usually been dated in or around 1680, and has been associated with the music for Nathaniel Lee's *Theodosius* Z606, first performed in that year.[17] But the work is much larger and more sophisticated than Z771, and its five-part scoring, for two violins, two violas, and bass, suggests a rather later date. The scoring was common in Italy—it is found in works by Giovanni Legrenzi, Giovanni Battista Vitali, Giovanni Bononcini, Antonio Caldara, and others—but was unknown in England before 1687, when Giovanni Battista Draghi used it in his St Cecilia ode 'From harmony, from heavenly harmony'. Purcell used it in his Queen Mary birthday odes of 1689 and 1690, while Blow used it for the 1690 New Year ode 'With cheerful hearts let all appear'. Thus Z771 may have been written for a lost ode in 1687–90.

There are two pieces in Add. MS 30930 that were probably written for the Twenty-four Violins. The Suite in G major Z770 is obviously incomplete, since Purcell left a blank space between the fourth and fifth movements, apparently for a sixth, and only copied the outer parts of the third, fourth, and fifth movements, leaving the staves for the inner parts blank. The fourth, a minuet, was also used as the last movement of the suite for Settle's play *Distressed Innocence* Z577/8, first produced in October 1690, so Thurston Dart borrowed its neighbour, the gavotte-like air Z577/7, for his edition of Z770 in *Works II*, 31. Two other movements in Z770 are known elsewhere: the second is the same as an air in the suite for *The Gordian Knot Unty'd* Z597/4, a lost comedy probably produced in November 1690, while the last, a jig with the popular song 'Hey, boys, up go we' in the bass, appears as a ritornello in 'Ye tuneful Muses' Z344/5c, the 1686 welcome ode for James II (see Ch. 5).

17 Campbell, *Henry Purcell*, 60–1.

These concordances enable us to date Z770 between 1686 and 1690. I argue in Ch. 5 that Purcell used 'Hey, boys, up go we' in 'Ye tuneful Muses' as a message of support for James II, so it is unlikely that he used it in the suite before he used it in the ode. Also, he made a revealing change to the second movement. In the autograph of Z770/1c he added a revised version of the last few bars of the top part on a blank stave, apparently in an attempt to avoid two turns into the subdominant within a few bars. The latest version seems to be the one in *The Gordion Knot Unty'd* suite, for there Purcell kept the revised version of the top part, but altered the lower parts to fit (Ex. 3.2). The movements of Z770 are in the same key, so it does not seem to be incidental music for

Ex. 3.2. (*a*) Minuet from the Suite in G major Z770/3, bars 20–5, compared with (*b*) the revised version of violin 1, and (*c*) the Air from *The Gordion Knot Unty'd* Z597/4, bars 27–32

the theatre; it was probably intended to be used at court, though we have no means of knowing whether it was completed and performed.

The other orchestral piece in Add. MS 30930, the Chacony in G minor Z730, was probably written a few years earlier. The word 'chacony' is just a variant of the English 'chacone', the equivalent of the French *chaconne*, the Spanish *chacona*, and the Italian *ciaccona*. In fact, the title is something of a misnomer, for it is based on an elaboration of the four descending notes of the *passacaglia* (see Ch. 2). The two dances, distinct around 1600, became increasingly similar in the late seventeenth century, and composers began to confuse them. They still cause trouble today, as the convoluted attempts to distinguish them in musical dictionaries show. In Italy chaconnes and passacaglias were usually set as songs or for guitar, lute, or keyboard, but Lully wrote orchestral examples as early as 1658 for his ballet *Alcidiane*, and most English ones are in the consort repertory. The earliest example seems to be the three-part 'Chacone' by Robert Smith in GB-Och, Mus. MS 1183, fo. 9^{r}, and John Carr's *Tripla concordia* (1677), p. 30. There are none by Matthew Locke and John Banister, who wrote most of the court dance repertory in the 1660s and early 1670s; the pieces labelled 'chicona' and 'chiconae' by Locke in the second set of 'The Broken Consort' and in *Tripla concordia* do not use ground basses.[18]

We have no idea why Purcell wrote the Chacony, but it is certainly more suitable for dancing than some other chaconnes and passacaglias. For instance, Blow's four-part Chaconne in G major, probably written at about the same time, has elaborate fugal passages, florid divisions, and some complicated rhythmic patterns with shifting accents.[19] Moreover, its bass part divides occasionally, providing separate notes for a continuo instrument—a sign, we have seen, of domestic consort music. The Chacony, by contrast, is typical of orchestral dance music. It is noble and restrained in style, consisting largely of patterns of dotted notes (the prevailing rhythm of French chaconnes and passacaglias); there are no elaborate divisions, no fugal or canonic writing, no need for a continuo, and no sign that one was used—despite the impression given in *Works II*, 31.

[18] Locke, *Chamber Music II*, 48–9; id., *Suite in G from Tripla Concordia*, ed. P. Holman (London, 1980), 7.

[19] J. Blow, *Chaconne for String Orchestra*, ed. H. W. Shaw (London, 1958).

Indeed, the work is so powerful precisely because, as in the best of Locke's orchestral music, there is a creative tension between the conventions of the dance and an adventurous musical language. The eight-bar ground is certainly conventional in outline: it consists of the descending fourth of the *passacaglia* balanced by a cadence. But it is cunningly inflected—the second bar is unexpectedly F *sharp* and the fifth B *natural*—and it is harmonized with the greatest richness and freedom. The inner parts are as strong as the outer parts, and contribute most of the piquant dissonances. They are far removed in style from the bland *parties de remplissage* that French composers often farmed out to assistants, though, to judge from the markedly different shades of ink in the autograph, Purcell copied them later than the outer parts. It was standard practice in the sixteenth and seventeenth centuries to write dance music in two stages, the outer parts first; nevertheless, Purcell must have had the inner parts in mind in variations 6 and 14, when the ground migrates into the viola and the second violin (allowing a series of transitory modulations), and in 8 and 11, when the bass stops and the ground migrates respectively to the first violin and the viola. We have noted Purcell's fondness for what might be called free symmetry several times already. Here the paired events—two modulatory passages, two with the bass running in quavers, two with the bass silent, and so on—give the work a satisfying architectural quality.

It is instructive to compare the Chacony with a ground-bass piece written by Purcell in the tradition of contrapuntal consort music cultivated at court by the Broken Consort. Z731 was catalogued by Zimmerman as a fantasia, though it is just headed '3 parts upon a Ground' in R.M. 20.h.9, and it belongs to the genre known at the time as 'divisions on a ground' or just 'grounds'. In R.M. 20.h.9 it is written out in D major for three violins and bass, but it has the rubric 'playd 2 notes higher for F[lutes]', and there is an autograph fragment of the second treble part in F major and in the French violin clef (G1) in GB-Lbl, Add. MS 30932.[20] This led Thurston Dart to suggest that the piece was originally written for recorders, pointing out that in D the bass goes down to *B′*, below the range of the cello or the English six-string bass viol.[21] But the bass could have been played in D as it

[20] The F major version is in *Works II*, 31, 52–60, the D major in H. Purcell, *Fantasia Three Parts upon a Ground*, ed. D. Stevens and T. Dart (London, 1953).

[21] T. Dart, 'Purcell's Chamber Music', *PRMA* 85 (1958–9), 89.

stands on a bass violin or a great bass viol (a size larger than the ordinary bass viol), and that version certainly exploits the open strings of the violins in an effective way. Perhaps Purcell had both instruments in mind, much as collections were sometimes devised with double clefs and key signatures, so the music could be played in the treble clef on the violin, oboe, or flute, or in the French violin clef a third higher on the recorder.[22]

'Three Parts on a Ground' is as full of contrapuntal artifice as the Chacony is devoid of it. Purcell did not invent its six-note ground. Pachelbel wrote his famous canon in the same key on almost the same bass, probably around the same time, and probably by coincidence; more to the point, Christopher Simpson used it in his *Compendium of Practical Music* (1667) to show how canons might be introduced over a ground.[23] Purcell himself wrote in the 1694 edition of Playford's *Introduction to the Skill of Musick* that 'Composing upon a Ground' was 'a very easie thing to do, and requires but little Judgement', but added that 'to maintain Fuges upon it would be difficult, being confined like a Canon to a Plain Song'.[24]

Much of the division writing in Z731 is fugal or semi-canonic, and it contains four passages of strict canon. The first slips past before the listener has time to realize that the row of notes in the first treble is simultaneously inverted in the second and played backwards in the third. Purcell even transfers the ground into the upper parts at one point, using it as the subject of a four-part canon to which the bass, divided into two for the purpose, contributes (Ex. 3.3). Such things could easily seem sterile in the hands of a lesser composer, but Purcell uses the canons as foils to the variations of chaconne-like dotted rhythms and the passages of brilliant divisions, and he enlivens the simple progressions with

[22] The only English example of this practice known to me is a manuscript in the possession of Tim Crawford that contains the second treble part of an anonymous set of six suites for two trebles and bass dating from around 1700, five of which have double clefs and key signatures. However, the Walsh and Hare edition of William Corbett's *Six Sonatas with an Overture in 4 Parts* Op. 3 (1708) for trumpet, two violins, and continuo has the note 'all these Sonatas are to be Play'd w(i)th 3 Flutes [i.e. recorders] & a Bass in the French Key 3 notes Higher', and a number of late 17th-c. English solo sonatas exist in alternative versions for violin and recorder a third apart; see e.g. the inventory in D. Lasocki, 'The Detroit Recorder Manuscript (England, *c.*1700)', *American Recorder*, 23/3 (Aug. 1982), 95–102.

[23] J. Pachelbel, *Canon and Gigue*, ed. C. Bartlett (Wyton, 1990); C. Simpson, *Compendium of Practical Music*, ed. Lord, 87–8.

[24] Squire, 'Purcell as Theorist', 567.

harmonic surprises of every sort, culminating in the last variation, with its breathtaking side-slipping harmonies.

'Three Parts upon a Ground' has virtually no precedent in English consort music—divisions on a ground were traditionally for lute, keyboard, or solo bass viol—and the work seems to have

Ex. 3.3. 'Three Parts on a Ground' Z731, bars 93–122

Ex. 3.3. *cont.*

attracted only one imitation, the Ground in A minor for three violins and bass by Bartholomew Isaack (1661–1709) in Drexel MS 5061.[25] But the three-treble scoring connects it with the Broken Consort and its repertory of court consort music. The

[25] Holman, 'Bartholomew Isaack', 383.

group had only two violinists until the great German violinist Thomas Baltzar joined it in the summer of 1661. Three-violin music, a popular genre in northern Europe at the time, made its first appearance in English music with Baltzar's ten-movement Suite in C.[26] (The expatriate composer William Young published some three-violin sonatas at Innsbruck in 1653, but there is no sign that they were known in England; he died there in 1662, so the William Young who served at the Restoration court was a different man.[27]) Baltzar's suite seems to have inspired John Jenkins to write a set of ten fantasia suites for three violins, bass viol, and continuo, and Nicola Matteis, the Italian virtuoso who took London by storm in the 1670s and 1680s, wrote a remarkable set of divisions in D minor for the same combination; it was probably known to Purcell, for it is in Osborn MS 515. There is unlikely to have been a group at court still playing three-violin music in the late 1670s, when Purcell probably wrote 'Three Parts upon a Ground'. He may just have been attracted to the scoring by studying the earlier repertory.

There is another piece by Purcell for three violins and continuo, the Pavan in G minor Z752; it is next to the Chacony in Add. MS 30930, and there is a copy in Drexel MS 5061. Dart suggested that it was written in 1677 in memory of Matthew Locke—pavans often served as memorials in Tudor and early Stuart times—but its style suggests that Purcell had John Jenkins (d. 1678) in mind rather than Locke, though there is no evidence that it is an elegy for anyone.[28] Like a number of consort pieces by Purcell, it harks back to early seventeenth-century models. It is typical of the pavan in its Jacobean heyday in that it is on a large scale, using a minim beat, with three sections or strains of 10½, 9½, and 12 breves. The strains are sharply contrasted, and each is based on a different 'topic'. The second, for instance, is largely taken up with rising and falling chromatic lines, while the third consists of a single point of counterpoint that works the subject and its inversion simultaneously, subjecting it to subtle melodic and rhythmic changes in the process—a favourite device of Jenkins. The harmonic idiom is also closer to the plain but

[26] For Baltzar and English three-violin music, see P. Holman, 'Thomas Baltzar (?1631–1663), the "Incomperable Lubicer on the Violin"', *Chelys*, 13 (1984), 19–21; id., *Four and Twenty Fiddlers*, 276–80.

[27] W. Young, *Sonate à 3. 4. e 5* (Innsbruck, 1653); ed. H. and O. Wessely (DTÖ 135; Graz, 1983).

[28] Dart, 'Purcell's Chamber Music', 89.

purposeful style of Jenkins than the more highly coloured style of Locke.

Jenkins was the last composer before Purcell to write large-scale pavans of the older type; there are two fine examples in his Lyra Consorts, written probably in the 1650s.[29] By the 1660s a more modest type of pavan had developed, exemplified by those that open a number of Locke's suites, including the ones in the second set of 'The Broken Consort'.[30] They still have three strains, but they are much shorter, and are based on the more modern crotchet beat, so they often sound more like solemn almands than true pavans. A single mood tends to prevail, and the interest is created less by the counterpoint than by the angular and unexpected melodic and harmonic idiom.

Purcell's four pavans for two violins and bass, the G minor Z751, the A minor Z749, the A major Z748, and the B flat major Z750, belong to this type, though they do not, as far as is known, come from suites. Indeed, they form an orderly sequence in the only source, GB-Lbl, Add. MS 33236, rising from *G* or gamut, and it may be that he planned to create a complete series of three-part pavans in every common key, but never finished the project. All the pavans by Locke and his contemporaries are placed at the head of suites, and we have to go back to the reign of Charles I—to the pavans in William Lawes's Harp Consorts or to Jenkins's five- and six-part viol consorts—to find a similar number of independent examples.

The same can be said of Purcell's fantasias. They have often been connected with Locke's four-part fantasias, but Locke always used them to open suites, and virtually all fantasias written after about 1640 are attached either to a single dance (as in Jenkins's fantasy–air pairs), two (as in the fantasia suite), or three (as in many of Locke's suites). Purcell's fantasias, by contrast, appear as single items in Add. MS 30930, grouped according to the number of parts: three fantasias in three parts Z732–4, nine in four parts Z735–743 (dated between 10 June and 31 August 1680), the opening section of a tenth four-part piece Z744 (24 February 1683), the five-part Fantasia upon One Note Z745, and two In Nomines, in six parts Z746 and seven parts Z747. Purcell's plan, comparable to that used by Locke in his autograph score GB-Lbl, Add. MS

[29] J. Jenkins, *The Lyra Viol Consorts*, ed. F. Traficante (RRMBE 67–8; Madison, Wis., 1992), 39–44, 58–63.

[30] Locke, *Chamber Music II*, 31–56.

17801, was evidently to assemble a collection of fantasias in an orderly sequence from the smallest number of parts to the largest. But the scheme was never completed, to judge from the number of blank pages in Add. MS 30930, and the headings 'Here Begineth the 5 Part: Fantazies' and 'Here Begineth the 6, 7 & 8 part Fantazia's'.

It is not clear why Purcell wrote them. Today they are a valued part of the viol repertory, but viols were dropping out of use before Purcell was born. Anthony à Wood chronicled the change to violins among Oxford amateurs in the late 1650s, and one wonders whether Purcell would have been able to assemble a complete viol consort from members of his circle around 1680, though the bass viol remained in use as a solo instrument and for continuo.[31] Roger North (who knew Purcell and played with him on several occasions; see below) thought Locke's Consort of Four Parts the last viol consort music: 'after Mr Jenkens I know but one poderose consort of that kind composed, which was Mr M. Lock's 4 parts, worthy to bring up the 'rere, after which wee are to expect no more of that style'.[32]

It used to be thought that Purcell wrote his fantasias for violins, or for mixed consorts with violins taking the upper parts. But there is no sign of the dance-like idiom used in contrapuntal music with violins ever since the reign of James I—there are no passages of triple time, for instance—and Purcell nearly always conforms to the part-ranges and types of writing traditionally used in the viol repertory. The top part of Z740 does go up to *d'''*, but this note is as unusual in violin music as in viol music; *c'''*, the fourth-finger extension in first position on the *e''* string, was the normal limit of consort violin parts, in England as on the Continent.

The most likely explanation is that Purcell wrote his fantasias more as composition exercises than as material for performance. Only Z733 survives in parts as well as score (in GB-Lbl, Add. MS 31435, with fantasias by Locke from the Consort of Four Parts, and fantasia suites by Christopher Gibbons), and they did not circulate outside Purcell's immediate circle, to judge from the surviving sources, even though items from the earlier viol

[31] For the situation in Oxford, see Holman, *Four and Twenty Fiddlers*, 267–75.

[32] Wilson, *Roger North*, 301; the Consort of Four Parts is in Locke, *Chamber Music II*, 57–97.

repertory were still being copied at the time.[33] All the complete four-part fantasias were written in a few weeks in the summer of 1680, perhaps as part of an intensive programme of study devoted to mastering contrapuntal techniques.

In some cases the objects of his study are known. Purcell used a fragment of a wordless score in his hand of Monteverdi's madrigal 'Cruda Armarilli' as a patch for the autograph of the Benedicite in B flat Z230M/3 in GB-Ob, MS Mus. A.1. There are madrigals by Monteverdi, Marenzio, and other Italian composers in the Jacobean and Caroline consort repertory; some have been identified as the models for fantasias by Ferrabosco, Coprario, and others.[34] A manuscript formerly in the possession of Thurston Dart contains a copy in Purcell's hand of a ten-part canon by John Bull on the Miserere plainsong, as well as organ parts for fantasias and fantasia suites by Orlando Gibbons and Coprario, and several madrigals by Monteverdi—including 'Cruda Armarilli'.[35] Osborn MS 515 includes the bass part of five-part viol consort suites by William Lawes as well as two sets by Jenkins (for treble, bass, and two lyra viols, and treble and two basses) that unfortunately do not survive elsewhere.

Purcell's models tended to vary according to the number of parts he was writing in. The classics of the three-part repertory were the nine fantasias by Orlando Gibbons, printed around 1620.[36] Their influence can be traced in most subsequent three-part fantasias, and it is certainly strong in Z732. The piece is relatively short—only sixty-one four-crotchet bars—and has a rapid succession of busy contrapuntal points, several of which use lively syncopated ideas. It is only when the music plunges unexpectedly into C minor and F minor in the middle, or when the counterpoint gives way to a heartfelt concluding passage of chromatic harmony marked 'Drag', that it becomes apparent that it is the work of a Restoration composer.

The other two, Z733 and Z734, are rather more ambitious and sophisticated, though Z733 went through several stages before the

[33] This point is discussed at length in Thompson, 'English Music Manuscripts', ch. 9.

[34] F. B. Zimmerman, 'Purcell and Monteverdi', *MT* 99 (1958), 368–9; see also C. Monson, *Voices and Viols in England, 1600–1650: The Sources and their Music* (Ann Arbor, 1982); J. Wess, '*Musica Transalpina*, Parody, and the Emerging Jacobean Viol Fantasia', *Chelys*, 15 (1986), 3–25.

[35] T. Dart, 'Purcell and Bull', *MT* 104 (1963), 30–1.

[36] O. Gibbons, *Consort Music*, ed. J. Harper (MB 48; London, 1982), nos. 7–15; see also Holman, *Four and Twenty Fiddlers*, 218, 220–2.

version in Add. MS 30930 (the one normally heard today) was arrived at. The earliest, in Add. MS 31435, consists of only fifty-two four-crotchet bars. Purcell probably added the fine chromatic conclusion, which follows an F major cadence with an E major chord and then gradually returns home through a maze of chromatic inflections, after studying the second D minor fantasia in Locke's Consort of Four Parts, which ends with a similar surprise.[37] Z734 is no less remarkable, though it owes less to Gibbons. The first half is a single point of counterpoint, taking a sombre theme through a series of subtle changes in the manner of Jenkins. After a perfect and seemingly final cadence there is a complete change, from minims and crotchets to crotchets and quavers, and from grave counterpoint to bustling homophonic music. All is not what it seems, however, for the passage is actually in invertible counterpoint, and the rest of the piece is taken up with working the three themes in every conceivable combination (Ex. 3.4). Purcell could not resist such feats of ingenuity in his youth, though he never allowed them to obscure the emotional message of the music.

Purcell's nine four-part fantasias are closest in formal outline to the fantasias of Locke's Consort of Four Parts. They are on a larger scale than the three-part fantasias, and have more sharply defined and contrasted sections. Nos. 5 and 9 Z736 and Z740 (11 and 23 June 1680) begin with passages that act as slow introductions and are marked off with a double bar (which may indicate a repeat), as in a number of Locke's fantasias, including nos. 1, 3, and 5 of the Consort of Four Parts. All the four-part fantasias except no. 12 Z743 (31 August 1680) have brief chordal passages in the middle, sometimes marked 'slow' or 'drag', that act as an interlude between the sections of busy counterpoint, and nos. 6 and 8 Z737 and Z739 (14 and 19/22 June 1680) also end with a passage marked 'slow'. The resulting patterns, effectively fast–slow–fast, slow–fast–slow–fast, or fast–slow–fast–slow, are the ones mostly found in Locke, though it must be said that most of the tempo changes in the modern editions of Locke have been added by his editors. Purcell's tempo marks nearly always accompany changes to faster or slower note-values, so it may be that they should be thought of as descriptive rather than prescriptive,

[37] Locke, *Chamber Music II*, 64–7.

Ex. 3.4. Fantasia in G minor Z734, bars 38–73

Ex. 3.4. *cont.*

and that the fantasias of both composers should be played at more or less a consistent pulse.[38]

However, Purcell's four-part fantasias are far more varied and adventurous than Locke's in their musical language and contrapuntal technique. Locke's harmony and part-writing is often delightfully angular and wayward, but there is nothing in the Consort of Four Parts to match the long-range harmonic planning of no. 4 Z735 (10 June 1680) in G minor. Here the flatwards drift of the first section—to C minor and F minor—is balanced by a sudden transition to sharps in the central slow section, touching on E minor, B minor, F♯ minor, A major, D major, C major, E minor, B minor, and D minor in seven four-minim bars. Similarly, the 'Brisk' section of Z739 touches on ten keys (D minor, A minor, G major, C minor, B flat major, E flat major, F minor, A flat major, B flat major, G major, E minor, A minor, and C major) in as many four-minim bars.

The model for harmonic plans of this sort was Jenkins rather than Locke. Jenkins was more interested in smooth, wide-ranging modulations than sudden surprising progressions, and several of his fantasias range right round the key cycle, involving enharmonic transitions.[39] Purcell never went that far (though D♯s, G♯s, and F♯s follow hard on the heels of E♭s, A♭s, and G♭s at one point in Z739), and in fact the speed with which he modulates tends to make the music sound like Locke even when he is following a Jenkins-like harmonic plan. Locke often raised the emotional temperature of his consort music by presenting relatively conventional harmonic events at a frenetic pace.

It is in his contrapuntal technique that Purcell departs furthest from Locke. The imitative points in the fantasias of the Consort of Four Parts are mostly based on one or two ideas; only in Fantasia no. 5 does Locke present three ideas simultaneously, and nowhere does he use inversion or augmentation. Inversion is found in most of Purcell's four-part fantasias, and it is combined with augmentation in the opening section of Z739, with single and double augmentation in the opening section of Z735, and with single, double, and triple augmentation in an astonishing passage towards the end of no. 12 Z743 (31 August 1680) (Ex. 3.5). Such

[38] E. TeSelle Boal, 'Purcell's Clock Tempos and the Fantasias', *JVGSA* 20 (1983), 24–39.

[39] See e.g. J. Jenkins, *Consort Music of Four Parts*, ed. A. Ashbee (MB 26; London, 1969), nos. 41, 44.

things had not been a regular part of English consort music for generations, and one has to go back to the reign of James I—to the canons of John Bull and Elway Bevin (praised by Purcell in *An Introduction to the Skill of Musick*, and copied by Daniel Henstridge in Add. MS 30933, fos. 141–61)—before encountering

Ex. 3.5. Fantasia in D minor Z743, bars 58–99

Ex. 3.5. *cont.*

a body of English music so taken up with formal contrapuntal devices.

In several four-part fantasias Purcell used recurring motifs to link two or more of the sections. In no. 11 Z742 (16, 18, 19 August 1680) the opening idea, heard rising and falling in inver-

Ex. 3.5. *cont.*

sion at the beginning, is used in its falling form in conjunction with two new ideas in the third 'brisk' section. In Z736 and Z740 an idea heard in the introduction is taken up in the succeeding contrapuntal point, and in Z740 one of the other themes of that point is transformed in turn into the theme of the next section. In

Z739 a rising or falling fourth is present in the subjects of all four main sections. English composers usually constructed their fantasias as a series of unrelated and sharply contrasted points, like madrigals, though monothematic works are not unknown: there are some by Jenkins that subject a single idea to a series of subtle melodic and rhythmic changes.[40] Also, Purcell might have been aware of Frescobaldi's canzonas, where similar devices are used; Frescobaldi often occurs in English keyboard manuscripts, and Blow incorporated several passages from the toccatas into his organ voluntaries.[41]

The six- and seven-part In Nomines Z746 and Z747 show that Purcell did not restrict his studies to seventeenth-century consort music. They belong to a genre that had its origin in John Taverner's mass 'Gloria tibi Trinitas', written probably in the 1520s.[42] A passage in the Benedictus at the words 'In nomine Domini' was detached from the mass, and became the model for more than 150 instrumental pieces written between the 1550s and the 1640s using the 'Gloria tibi Trinitas' plainsong as a cantus firmus. Purcell was almost certainly the only composer to write In Nomines after then: the six-part examples by Jenkins were probably written in the 1630s, while the five-part 'In Nomine Fantasia' by 'J.B.' in GB-Ob, MSS Mus. Sch. 473–8, ascribed by Dart and others to John Banister, seems to be by the parliamentary official John Browne (1608–91), and was written in imitation of one of Lawes's six-part In Nomines, probably in the 1640s.[43]

In many respects, Z746 and Z747 are closer in style to sixteenth- rather than seventeenth-century In Nomines. The only other examples of seven-part writing in the English viol repertory are two In Nomines by Robert Parsons (d. 1570), and a fragmentary one by Robert White (d. 1574) in GB-Lbl, Add. MS 32377.[44] In Z747 Purcell followed tradition in placing the cantus firmus in even breves in the alto part, and surrounded it with smooth polyphony, rising and falling in minims and crotchets. The edifice is articulated, with a sure sense of architecture, by two bars of

[40] A. Ashbee, *The Harmonious Musick of John Jenkins*, i: *The Fantasias for Viols* (Surbiton, 1992), 212–43.

[41] G. Cox, *Organ Music in Restoration England: A Study of Sources, Styles, and Influences* (New York and London, 1989), i. 140–9, 172–205.

[42] C. Hand, *John Taverner: His Life and Music* (London, 1978), 46–52.

[43] Dart, 'Purcell's Chamber Music', 90; D. Pinto, 'William Lawes's Music for Viol Consort', *EM* 6 (1978), 14–15, 24.

[44] P. Doe (ed.), *Elizabethan Consort Music I* (MB 44; London, 1979), nos. 74, 75.

rich seven-part harmony, placed exactly at the midway point. We are a long way here from the busy writing in quavers and semi-quavers of In Nomines by Ferrabosco, Gibbons, and Jenkins, or the sharp fantasy-like contrasts of those by Lawes; indeed, only a delicious slide into B flat minor towards the end suddenly makes the listener aware that this is music from around 1680, not 1580. Z746 sounds even more archaic, and uses a technique found in early sixteenth-century motets: the cantus firmus provides the material for all the counterpoint, and is speeded up so that it can be heard as a tune. There is no precedent for the playful Fantasia upon One Note Z745, in which a cantus firmus is reduced *ad absurdum* to a single middle C—the object being, evidently, to demonstrate that even the severest limitation need not restrict the imagination.

Purcell's interest in formal counterpoint did not cease when he stopped writing fantasias. It also seems to have been one of the reasons—perhaps the main one—why he began to study and imitate Italian trio sonatas. In the sonata, 'the chiefest Instrumental Musick now in request', he wrote in *An Introduction to the Skill of Musick*, 'you will find Double and Treble Fuges also reverted and augmented in their Canzona's, with a good deal of Art mixed with good Air, which is the Perfection of a Master'.[45] Most of Purcell's trio sonatas come from the early 1680s. Twelve, Z790–801, were published by the author in an engraved edition, *Sonnata's of III Parts* (London, 1683). The rest, Z802–11, appeared in *Ten Sonata's in Four Parts* (London, 1697), though eight of them are in Add. MS 30930 in whole or part, and the first three seem to have been copied before 1684, for the tempo marks have Purcell's early reversed type of 'r', which disappears after 1683 in the chronological sequence of pieces in GB-Lbl, R.M. 20.h.8.[46] The more modern 'r' is found in the copy of Sonatas 7, 8, and 9, and so they are probably a little later; no. 10, Purcell's most consistently modern trio sonata, was probably copied last of all.

The scoring of the sonatas has caused much confusion. Despite the titles of the two sets, both consist of four part-books, and call for the same instruments: two violins, bass viol, and organ or harpsichord.[47] In describing the 1683 set as 'of three parts'

[45] Squire, 'Purcell as Theorist', 557–8.

[46] Thompson, 'Purcell's Great Autographs'; see also D. Stevens, 'Purcell's Art of Fantasia', *ML* 33 (1952), 341–5.

[47] The 1683 title-page specifies 'TWO VIOLLINS And BASSE: To the Organ or

Purcell was following Italian practice, which did not count the continuo as a separate part; in the same way, Corelli's trio sonatas were said to be 'à tre' though they are printed in four part-books. It also seems that Purcell did not intend originally to publish a separate continuo part. According to the preface, the publication 'had been abroad in the world much sooner, but that he has now thought fit to cause the whole Thorough Bass to be Engraven, which was a thing quite besides his first Resolutions'. He may have intended to publish just the string parts, as Gibbons had done with his three-part fantasias, expecting the organist to read from a manuscript score or a keyboard reduction. Or he may have tried to combine the continuo and the bass on one stave; if so, it would have proved impractical, so often do they diverge. The 1697 set was presumably said to be 'of four parts' because there were four part-books. This was the rule in eighteenth-century England: Handel's Op. 6 concertos, for instance, were described on the title-page as 'in SEVEN PARTS' because there were seven part-books; the music is almost never in seven real parts.

Purcell's sonatas do not seem to have been written for the court. Roger North wrote that when he first lived in London, in the early 1670s, he was one of 'that company which introduc't the Itallian composed enterteinements of musick which they call Sonnata's'.[48] At the time the fashion at court was only for 'the theatricall musick and French air in song', but the Norths and their friends 'found most satisfaction in the Italian, for their measures were just and quick, set off with wonderful solemne Grave's, and full of variety'. Purcell certainly moved in such circles, for North remembered that his brother Francis, the Lord Chief Justice, 'caused the devine Purcell to bring his Itallian manner'd compositions; and with him on his harpsichord, my self and another violin, wee performed them more than once, of which Mr Purcell was not a little proud, nor was it a common thing for one of his dignity to be so enterteined'.[49] Incidentally, it is unlikely that Purcell's sonatas were performed in the Chapel Royal, as has been suggested. Sonatas were used in the liturgy in Italy and parts of Catholic Europe north of the Alps, but there is no evidence

Harpsecord', though an advertisement in *Choice Ayres*, v (1684), 63 describes the scoring more specifically as 'two Violins and Bass-Viol, with a Through-Bass for the Organ or Harpsichord'; the parts of the 1697 set are labelled 'VIOLINO PRIMO', 'VIOLINO SECUNDO', 'BASSUS', and 'Through Bass for the Harpsichord, or Organ'.

[48] Wilson, *Roger North*, 25.

[49] Ibid. 47.

that the same was true of Anglican churches and chapels, even at court. The string-players who served in the Chapel Royal were there, it seems, to play the string parts of anthems, not sonatas.

According to the famous preface of the 1683 set, Purcell 'faithfully endeavour'd a just imitation of the most fam'd Italian Masters' in his sonatas, and much ink has been split trying to identify them. Roger North wrote that 'severall litle printed consorts came over from Italy, as Cazzati, Vitali, and other lesser scrapps which were made use of in corners', and English manuscripts do contain many Italian sonatas; a good example is GB-Lbl, Add. MS 31436, copied by someone in the North circle, which contains items from G. B. Vitali's Op. 9 (Venice, 1684) and the trio sonata anthology *Scielta delle suonate* (Bologna, 1680).[50] The problem is that most manuscripts cannot be dated exactly, so we do not know, for instance, whether Corelli's Op. 1 (Rome, 1681) arrived in time for Purcell to take note of it. An exception is GB-Lbl, Add. MS 31431, dated 1680 and owned by Sir Gabriel Roberts (*c.*1630–1715), which includes sonatas from Legrenzi's Op. 2 (Venice, 1655), Cazzati's Op. 18 (Venice, 1656), and Vitali's Opp. 2 and 5 (Bologna, 1667 and 1669).[51]

To judge from the number of surviving copies, the Italian instrumental music most admired by English musicians was a group of unpublished works known to them as sonatas by the Roman lutenist Lelio Colista; they exist in at least seven English manuscripts, and Purcell printed a passage of one of them in *An Introduction to the Skill of Musick* as an example of invertible counterpoint or 'Double Descant'.[52] In fact, they are labelled 'simfonia' rather than 'sonata' in the main Italian source, and about half of them—including the one quoted by Purcell—are actually by the Milanese violinist Carlo Ambrogio Lonati, who worked in Rome from about 1668 to 1677, and supposedly visited England in the later 1680s.[53]

The relationship between Purcell's sonatas and those by the 'fam'd Italian masters' is complex and not yet fully understood. It is further complicated by the existence of a number of trio sonatas

[50] Ibid. 302; P. Holman, 'Suites by Jenkins Rediscovered', *EM* 6 (1978), 26; Thompson, 'English Music Manuscripts', 325.

[51] Ibid. 377–80, 436–43.

[52] Squire, 'Purcell as Theorist', 557.

[53] Thompson, 'English Music Manuscripts', 450–3; see also P. Allsop, 'Problems of Ascription in the Roman *Simfonia* of the Late Seventeenth Century: Colista and Lonati', *MR* 50 (1989), 34–44; id., *The Italian 'Trio' Sonata from its Origins until Corelli* (Oxford, 1992), 192–9.

that might or might not have been written in England before Purcell's, such as Blow's in A major (in R.M. 20.h.9, Add. MS 33236, and several Oxford sources), Draghi's in G minor (Add. MS 33236), Nicola Matteis's in A major (GB-Ob, MSS Mus. Sch. E. 400–3), Robert King's in A major (GB-Ob, MSS Mus. Sch. E. 443–6), and Isaac Blackwell's in F minor (Add. MS 31431). By and large, however, Purcell seems to have looked to the works of the older generation of Italians for his models, such as Cazzati (b. *c.* 1620), Legrenzi (b. 1626), and Colista (b. 1629), rather than those of his near contemporaries, such as Corelli (b. 1653) and Bassani (b. *c.* 1657). His sonatas are 'à tre' rather than 'à due', which means that the bass viol contributes to the musical argument on more or less equal terms with the violins, often parting company with the continuo. The sonata à tre was a conservative contrapuntal form, descended from the early seventeenth-century canzona, while the more forward-looking sonata à due, with its purely harmonic continuo (normally played in Italy at the time without the support of a melody instrument), was closer in style to secular vocal music and dance music.[54] There is certainly no sign in Purcell's sonatas of the slightly later distinction, formalized by Corelli, between the *da chiesa* and *da camera* types: dances are mixed freely with 'abstract' movements, as in mid-century sonatas.

Purcell mostly used formal patterns established long before Corelli. His sonatas tend to consist of five or more short linked sections rather than the more modern sequence of four discrete movements, the type that predominates in Corelli's Op. 1. There are a few thematic relationships between movements, a common device earlier in the century, and they sometimes effect a recapitulation: the fugal opening themes of the sonatas in A minor and G major Z794 and Z797, 1683 nos. 5 and 8, both appear near the end combined with new material, a device found in a number of the sonatas in Legrenzi's Op. 2.[55] The slow movements often use a common Italian pattern in which a short opening phrase is followed after a pause by a repetition in the dominant, which is then extended by modulations through a number of keys. The fast movements are either essays in the canzona style—whether or not

[54] Allsop, *The Italian 'Trio' Sonata*, esp. ch. 3; id., 'The Role of the Stringed Bass as a Continuo Instrument in Italian Seventeenth-Century Instrumental Music', *Chelys*, 8 (1978–9), 31–7.

[55] G. Legrenzi, *Sonate a Due a Tre Opus 2 1655*, ed. S. Bonta (Harvard Publications in Music, 14; Cambridge, Mass., 1984), nos. 1, 3, 5, 10–13, 15.

they are actually labelled 'canzona'—or triple-time dances. There is only one example, the Vivace of the Sonata in G minor Z809, 1697 no. 8, of the type of running-bass movement familiar to us from Corelli, and none of the driving non-fugal allegro we think of as the epitome of the Italian Baroque style. This sort of movement was originally associated with the sonata à due, but it crept into the sonata à tre in the 1680s; there are early examples in Corelli's Op. 1 and Bassani's Op. 5 (Bologna, 1683).

The canzonas of Purcell's sonatas are closely modelled on those in the Colista–Lonati repertory. A common type presents two themes at the outset in invertible counterpoint, one striding in minims and crotchets, the other running in quavers and semiquavers; they are present in some form throughout, and the object of the exercise is to keep up the interest by combining them in new ways, in inversion, augmentation, and stretto. In another type, common in the fugues of Purcell's late overtures, the themes are presented successively, the second arriving just as the ear tires of the first. In some cases, as in the canzona of the Sonata in A minor Z804, 1697 no. 3, they are combined, while in others, such as the jig-like movement that succeeds it, the second supplants the first. In a related type, also found in the à tre sonatas of Legrenzi's Op. 2, the canzona emerges out of, and is effectively the second section of, a galliard-like movement in 3/2, as in the Sonata in G major Z797, 1683 no. 8, or a minuet-like movement in 3/4, as in the Sonata in D major Z811, 1697 no. 10. The first section is always marked 'largo' or 'poco largo' and the second 'allegro' or 'vivace', but it is unlikely that he intended much change of pace, given that 'largo' was 'a middle movement' in tempo, according to the preface of the 1683 sonatas. Indeed, these double-barrelled movements are most effective when their contrasted elements—simple dance and elaborate counterpoint—are united in a single tempo.

The rarest type of canzona in Purcell's sonatas is the one where a single theme predominates, but gives way at regular intervals to freer episodes. This, of course, was to be the standard pattern of the eighteenth-century fugue, and it is significant that the best example is in Z811, possibly Purcell's last trio sonata. The triadic theme, in the fashionable D major fanfare style of Italian trumpet music, is matched by unusually simple harmonies, see-sawing regularly between tonic and dominant before being swept into the dominant proper by an episode of rising and falling sequences (Ex. 3.6).

Ex. 3.6. Sonata in D major Z811, bars 36–46

Ex. 3.6. *cont.*

It is easy to be beguiled by the noble simplicity of this music into missing its subtleties: how naturally a fragment of the countersubject provides, in inversion where necessary, all the material for the episode; and how easily and casually the subject itself is inverted on its return.

The sequence of sonatas in the two collections requires some comment. It was common practice during the Restoration period to organize collections of instrumental music in an ascending sequence of keys from gamut (G major and minor), and sets of fantasia suites were often laid out with the works in pairs; the two by Lawes, for instance, use the same sequence of keys: G minor, G major, A minor, C major, D minor, D major, D minor, and D major. Purcell uses a more complex sequence in the 1683 set. The first eight sonatas ascend from gamut, but in the arpeggiated pattern G minor, B flat major, D minor, F major, A minor, C major, E minor, and G major, so that major and minor works alternate. The last four, C minor, A major, F minor, and D major, make up a falling arpeggio, balancing the rising one.

Purcell obviously chose the order of the 1683 sonatas, but that is not necessarily true of the 1697 sonatas. It is odd, for instance, that there are ten works, rather than six or twelve, the most common number of sonatas or concertos in Continental publications, or eight, a common number for sets of fantasia suites. There have been a number of attempts to explain the apparently random sequence B minor, E flat major, A minor, D minor, G minor, G minor, C major, G minor, F major, and D major, but it is difficult to understand why Purcell included three works in G minor if he intended a structured pattern like the 1683 set. However, it may not be chance that the ten works can be rearranged to make a rising sequence from gamut: nos. 8, 3, 1, 7, 4, 10, 2, 9, 5, and 6. No. 6, the great set of variations on the same five-bar *passacaglia*-like ground as Lully's 'Scocca pur, tutti tuoi strali' (see Ch. 2), belongs naturally at the end of the set, just as Georg Muffat's *Armonico Tributo* (Salzburg, 1682) ends with a massive passacaglia, or Corelli's Op. 2 (Rome, 1685) ends with a *ciaccona*.

There is only one Purcell trio sonata outside the two published sets: Z780. This used to be known as the 'Violin Sonata in G minor' because the only known source, copied by the York cleric Edward Finch (1663–1738) into a missing manuscript last seen at Sotheby's in 1935, is or was on two staves, for violin and continuo; fortunately, the work was published several times around 1900 in versions for violin and piano.[56] Thurston Dart noticed that in the fast movements there are a number of imitative entries in the bass missing, and that it was possible to add a bass viol part that restored them by ornamenting the simple continuo line. He therefore published in *Works II*, 31 his reconstruction of a hypothetical original version for violin, obbligato bass viol, and organ continuo, along the lines of similar works in the English and German repertory. His version of the fast movements is extremely convincing, though he was on shakier ground in the homophonic slow movements; works for the same scoring by Purcell's colleagues and followers, such as Gottfried Finger's Op. 1/1–3 (1688), or William Corbett's Op. 1/1 (*c*.1700), suggest that the viol part should be rather more elaborate and chordal. The work, in four distinct movements, slow–fast–flow-fast, sounds more modern than all the other trio sonatas except Z811; it therefore probably dates from the later 1680s. Perhaps it was inspired by Finger's three examples, published in a collection dedicated to James II and containing, according to the title-page, works played in his Catholic chapel.

Purcell's sets of sonatas do not seem to have been very successful. They circulated more widely than the rest of his consort music because they were printed, but they did not sell well.

[56] See *Works II*, 31, p. xii for a facsimile; for its history, see R. Illing, *Henry Purcell: Sonata in G Minor for Violin and Continuo: An Account of its Survival from both the Historical and Technical Points of View* (Flinders University, South Australia, 1975); for Edward Finch and his copying activities, see D. Nichols, 'Edward Finch and the Associated Influential Composers and their Music' (BA diss., Colchester Institute, 1989).

Frances Purcell extended the time-limit for subscriptions to the second set in 1696 and reduced the price in 1699; there were copies of both sets left unsold at her death in 1706.[57] The problem was partly the familiar one that the public saw no reason to prefer an imitation to the genuine article. Corelli's trio sonatas circulated in England in manuscript in the 1680s, and were all readily available in Antwerp reprints by the early 1690s. 'Then came over Corelly's first consort', Roger North wrote, 'that cleared the ground of all sorts of other musick whatsoever.'[58] North thought Purcell's 'noble set of sonnatas' 'very artificiall and good musick', but admitted they were 'clog'd with somewhat of an English vein, for which they are unworthily despised'.

Today taste has come full circle. We tend to like music clogged with the English vein, such as the Sonata in C minor Z798, 1683 no. 9. But English musicians in the 1680s and 1690s had good reason to welcome the clarity, order, and logic of the Corelli style; they were tired of the perpetual surprise of earlier English music as well as the lack of seriousness of the French style—'the levity, and balladry of our neighbours', as the preface to the 1683 set put it. In particular, Corelli's sequential patterns and logical modulation schemes were obvious models for a generation of English composers trying to clarify and expand the horizons of their harmonic thinking. It is likely, too, that Purcell's sonatas suffered because they were serious contrapuntal works, intended to appeal to the player rather than the listener. During the 1690s the English public became accustomed to listening to instrumental music in concerts and in the theatre rather than playing it at home. More and more consort collections were published in London, but they tended to contain recycled concert and theatre works rather than genuine chamber music.

Purcell's harpsichord music was rather more successful than his sonatas, though no one would claim it was equal in importance.[59]

[57] M. Tilmouth, 'The Technique and Forms of Purcell's Sonatas', *ML* 40 (1959), 121; C. D. S. Field and M. Tilmouth, 'Consort Music II: From 1660' in Spink (ed.), *Seventeenth Century*, 275.

[58] Wilson, *Roger North*, 310–11.

[59] H. Purcell, *Harpsichord and Organ Music*, ed. W. B. Squire and E. J. Hopkins, *Works I*, 6 (London, 1895) has long been out of print; a new edition edited by C. Hogwood is in progress for *Works II*; meanwhile the best edition is Purcell, *Complete Harpsichord Works*, ed. H. Ferguson (EKM 21, 22; 2nd edn., London, 1968). For sources of Purcell's keyboard music discovered more recently, see R. Klakowich, 'Harpsichord Music by Purcell and Clarke in Los Angeles', *Journal of Musicology*, 4 (1985–6), 171–90, and the bibliography in n. 4; see also R. Charteris, 'Some Manuscript Discoveries of Henry Purcell and his

Part of the reason for its success was the rarity of English printed keyboard collections. *The Second Part of Musick's Hand-Maid* (1689) and *A Choice Collection of Lessons for the Harpsichord or Spinnet* (1696) were only the fifth and sixth ever printed in England, and the *Choice Collection* was the first to be confined to the work of a single composer; it was the model for similar collections by Blow, Clarke, and others. *The Second Part of Musick's Hand-Maid*, the sequel to a collection published by John Playford in 1663 and again in 1678, was reprinted in 1705, while *A Choice Collection* achieved at least two reprints, in 1699 and 1700.[60]

It is not clear how many pieces in *The Second Part of Musick's Hand-Maid* were written or arranged by Purcell. Eleven are attributed to him, some of which are simple arrangements of songs or sections from larger vocal works, and seven more can be ascribed to him with the help of concordances. Purcell edited the volume—Henry Playford wrote in the introduction that it was 'carefully Revised and Corrected by the said Mr. Henry Purcell'—so some of the other pieces, such as the anonymous 'Ayre' no. 1 (a simple setting of the tune 'Ham-House, or Cherry-Garden'; see *Apollo's Banquet*, 5th edn., 1687, no. 93), could well have been arranged by him.

This certainly seems to have been so in the case of no. 20, a keyboard version of 'Scocca pur, tutti tuoi strali'. It is a beautiful arrangement, with carefully calculated *style brisé* harmonizations of the statements of the ground between the vocal phrases, and it is strikingly similar in style to two keyboard versions of vocal ground basses by Purcell that have been accepted as his own work: no. 17 in the 1689 book, 'A new Ground' ZT682, a version of 'Here the deities approve' from the 1683 St Cecilia ode 'Welcome to all the pleasures' Z339/3, and ZT681, an arrangement, surviving only in manuscript, of 'With him he brings the partner' from the 1686 ode 'Ye tuneful Muses' Z344/11a.

Until recently, in the absence of autographs, editors treated the 1689 and 1696 collections as the primary sources of Purcell's keyboard music. That situation has now changed with the discovery of a manuscript containing twenty-one pieces in Purcell's hand, including six hitherto unknown, three of which are part of variant

Contemporaries in the Newberry Library, Chicago', *Notes*, 37 (1980), 7–13; P. Holman, 'A New Source of Restoration Keyboard Music', *RMARC* 20 (1986–7), 53–7.

[60] T. Dart (ed.), *The Second Part of Musick's Hand-Maid* (EKM 10; 2nd edn., London, 1968).

versions of the Suites in A minor Z663 and C major Z665.[61] The source, which also contains a sequence of seventeen apparently autograph pieces by Draghi at the other end of the book, seems to have been used by Purcell for teaching towards the end of his life. The first few pieces are evidently intended for a beginner. The sequence opens with a rudimentary C major prelude, and the next two airs are hardly worthy of the composer; indeed, the minuet, no. 2, with its crude parallel octaves at the beginning, sounds like a composition by the pupil rather than the teacher.

The Purcell–Draghi manuscript throws interesting and unexpected light on Purcell's musical taste. The breadth of his interests is neatly reflected by the copies of the G major prelude by Orlando Gibbons from *Parthenia* (1612–13; repr. 1646, 1651, and 1655), and an arrangement of a hornpipe attributed to John Eccles in *Apollo's Banquet* (8th edn., 1701). It is also of interest for the keyboard versions it preserves of movements from Purcell's theatre suites. It authenticates several arrangements hitherto known only from secondary sources, and it provides five new authentic keyboard arrangements, of pieces from *The Fairy Queen* Z629/1b and 44a, *The Double Dealer* Z592/7, and *The Virtuous Wife* Z611/7 and 8.

It is impossible to tell in most cases whether Purcell had a hand in the arrangements that only survive in secondary manuscripts—Howard Ferguson decided 'to print in full any transcription that at least sounds effective on the keyboard'[62]—and one or two have been accepted too readily into the Purcell canon, to judge from their quality. It is hard to believe, for instance, that he had anything to do with the 'Chacone' in G minor ZT680, a clumsy arrangement of the four-part Curtain Tune from *Timon of Athens* Z632/20, printed in a group of miscellaneous pieces at the end of *A Choice Collection*. Nevertheless, the number of such arrangements is eloquent testimony to the impression these wonderful pieces made on his contemporaries.

We are on surer ground with the eight suites in *A Choice Collection*, though even here there are considerable differences between the various sources. The Purcell–Draghi manuscript has a version of Suite no. 4 in A minor Z663 with a fine jig instead of

[61] Sold at Sotheby's on 26 May 1994; see R. Morrison, 'Purcell's Notebook Revealed', *The Times* (17 Nov. 1993), 35; I am grateful to Lisa Cox and Robert Spencer for enabling me to see the manuscript.

[62] Purcell, *Complete Harpsichord Works*, ii. 37.

the saraband printed in the 1696 collection, and different preludes survive for it and the Suite no. 2 in G minor Z661 in secondary manuscripts. The situation is more complex in the case of the Suite in C major Z666. In the 1689 book the suite (catalogued by Zimmerman separately as Z665) starts with an 19-bar arpeggiated prelude, and has a simple almand, largely in two parts. In the Purcell–Draghi manuscript there is another prelude, also arpeggiated but only nine bars long, and a different, more complex almand; the rest of the suite is the same. This almand is retained in what appears to be the latest version, in the 1696 collection, but the arpeggiated preludes are replaced by a brilliant if rather vacuous contrapuntal invention in the Italian style. Purcell was not alone in changing his keyboard suites in this way: similar things are found in John Blow's keyboard music.[63]

We have no ideas when Purcell wrote the eight suites. It would be tempting to think that they are roughly in chronological order, for no. 1 in G major Z660 is by far the simplest, and the last two, no. 7 in D minor Z668 and no. 8 in F major Z669, are the finest and the most elaborate. Nos. 7 and 8 conclude respectively with arrangements of a hornpipe from *The Married Beau* Z603/3 (April 1694), and a minuet from *The Double Dealer* Z592/3 (October or November 1693), so they presumably come from the last few years of Purcell's life, at least in the 1696 versions. Furthermore, the noble almand in Z668, arguably the finest movement of the set, is subtitled 'Bell-barr'—a name that also occurs in connection with the song 'I love and I must' Z382 (late 1693; see Ch. 2). It has yet to be explained properly, but Bell Bar is the name of a hamlet near the south end of the park of Hatfield House in Hertfordshire, and it could be that Purcell wrote both pieces there in the winter of 1693–4. Of course, there are other possible explanations for the order of the eight suites. They are in ascending order from gamut—G major, G minor, G major, A minor, C major, D major, D minor, and F major—and they may be graded to some extent; Z660 certainly looks like a piece for teaching beginners, and it has served generations of aspiring keyboard-players well.

Purcell's suites may seem remarkably diverse by the standards of a Continental composer such as Froberger, but English com-

[63] H. W. Shaw, 'The Harpsichord Music of John Blow: A First Catalogue', in O. W. Neighbour (ed.), *Music and Bibliography: Essays in Honour of Alec Hyatt King* (London, 1980), 51–68.

posers never really settled on a standard number or sequence of movements for their keyboard suites. In fact, Purcell's are more regular than most: all but one (the 1689 version of no. 5) are either in three or four movements, all but one (no. 7) start with a prelude, and all but one (no. 6 Z667) include the pairing of almand and corant; four have the sequence prelude, almand, corant, and saraband, and another substitutes a minuet for the saraband. The dances divide into two types in style: the almands and corants use the elaborate *style brisé* textures that suggest polyphonic part-writing by the use of broken chord patterns, meshing like the cogs of a fine Restoration time-piece (Ex. 3.7). The

Ex. 3.7. Almand from the Suite in D major Z667, bars 1–7

lighter dances use a much simpler type of texture, derived from the large repertory of arrangements of popular tunes and simple consort music: the tune in the right hand rides high above two parts in the left hand.

Music of this sort tends to sound clumsy and insignificant on a large harpsichord, but comes to life on the harpsichord's smaller single-strung relatives, the virginal and the spinet. Harpsichords seem to have been relatively rare in Restoration England. Only one English example survives—a single-manual instrument with two 8′ stops made in 1683 by Charles Haward[64]—whereas we have a fair number of English rectangular virginals, ranging in date from 1641 to 1679. The bentside spinet, a type apparently introduced to England in the 1660s by its inventor Girolamo Zenti, superseded the oblong virginal in the 1670s and 1680s, and was made in large quantities in London by Charles and Thomas Haward, Stephen Keene, and others; it was the standard keyboard instrument in ordinary English homes for the next century. Frances Purcell bequeathed three keyboard instruments to her son Edward in 1706: 'the organ, the double spinnet [possibly an instrument with an extended compass rather than with two keyboards], the single spinnet'.[65] They were presumably the instruments her husband had used.

The organ in the Purcell household reminds us that the chamber organ was also a common domestic instrument in England, and was the normal accompaniment for consort music until the 1690s. A fair number of Restoration chamber organs survive in whole or part. Most are by anonymous makers, and were once ascribed to Bernard 'Father' Smith; they are typically softly voiced single-manual instruments, with predominantly wooden pipes, and four or five stops.[66] The Restoration was also a period of intensive church-organ building. Instruments destroyed or damaged during the Civil War had to be restored or replaced, and new organs were required for the churches that were built in London following the Great Fire, as well as those in London's expanding suburbs.

The Dallams, the leading family of organ-builders before the Civil War, still received a good deal of the work, but faced increasing competition from Renatus Harris (whose mother was a Dallam) and Smith, among others. Virtually none of their organs survive in original condition, but they were mostly instruments with between twelve and twenty stops, some with two keyboards, one for the main or 'great' organ, the other for the smaller 'chair' organ placed behind the organist's back, which hung over the edge of the organ gallery. There were no pedals or 16′ stops, so there was little distinction between music written for chamber or

[64] Now at Hovingham Hall in Yorkshire; see D. H. Boalch, *Makers of the Harpsichord and Clavichord 1440–1840* (2nd edn., Oxford, 1974), 65–6.

[65] Zimmerman, *Purcell: Life and Times*, 283.

[66] M. Wilson, *The English Chamber Organ: History and Development 1650–1850* (Oxford, 1968); D. Gwynn, 'The English Organ in Purcell's Lifetime', in M. Burden (ed.), *Performing the Music of Henry Purcell.*

church instruments, except that Restoration composers often wrote voluntaries for double organ, exploiting solo reed stops or the contrast between the choruses of the two keyboards.

We only have a few organ voluntaries by Purcell. Several pieces that have been attributed to him in the past are now known to be spurious: the Voluntary in C major ZD241 is probably by John Barrett, while the Verse in the Phrygian Mode ZS126 is by Nicolas Lebègue, and the Voluntary on the 100th Psalm Z721 is just as likely to be by Blow as by Purcell. A single-organ version of Z721 is ascribed to the former in John Stafford Smith's *Musica antiqua* (1812), who printed it from a lost source, while a double-organ version is ascribed to the latter in GB-Lbl, Add. MS 34695.[67] The same manuscript also contains the much-disputed Toccata in A major ZD229 with an ascription to Purcell.

Scholars have long been baffled by this fine piece, which has also been ascribed to Michelangelo Rossi and J. S. Bach, but Barry Cooper has pointed out that its harmonic style is too modern for Rossi (d. 1656), and that J. S. Bach was too young to have written a piece that occurs in manuscripts dated 1698 and 1702.[68] By the same token, one of Gloria Rose's candidates, Wilhelm Hieronymus Pachelbel (an attribution to 'W. H. P.' might have been misread as 'M. H. P.'—Mr Henry Purcell), is equally unlikely, for Johann Pachelbel's son was only born in 1686. Nevertheless, ZD229 is also unlike any other keyboard piece by Purcell, or any by his English colleagues, for that matter. To judge from its Germanic style, it could be by one of the German organists active in Restoration England.[69]

The remaining five pieces in the standard modern edition of Purcell's organ works need not detain us long.[70] Two, the Verse in F major Z716 and the Voluntary in C major Z717, are trifles. The former is based on a single contrapuntal idea, while the later is similar in size and shape to some of the little voluntaries by Locke in *Melothesia* (1673): a full-voiced passage gives way in the middle to a spritely point of imitation.[71] The Voluntary in G

[67] Cox, *Organ Music in Restoration England*, i. 259–61.

[68] G. Rose, 'Purcell, Michelangelo Rossi, and J. S. Bach: Problems of Authorship', *AcM* 40 (1968), 203–19; B. Cooper, 'Keyboard Music', in Spink (ed.), *Seventeenth Century*, 364–5; Cox, *Organ Music in Restoration England*, i. 117–18.

[69] Ibid. i. 21–34.

[70] H. Purcell, *Organ Works*, ed. H. McLean (2nd edn., London, 1967).

[71] M. Locke, *Seven Pieces (Voluntaries) from 'Melothesia' (1673)*, ed. G. Phillips (Tallis to Wesley, 6; London, 1957).

major Z720 uses the same design on a large scale; the first section is in the *durezze e ligature* (dissonances and ties) style that English organists derived from Frescobaldi's toccatas.[72] By far the most important of Purcell's organ works, however, is the Voluntary in D minor Z718 and its cousin, the Voluntary for Double Organ Z719. The two pieces are more or less the same for the first page, but then gradually diverge until, about halfway through, they go entirely separate ways. We cannot be sure which came first, or that Purcell was wholly responsible for both of them, but they are both striking pieces, with lavishly ornamented imitative passages alternating with bursts of florid passage-work (Ex. 3.8).

Ex. 3.8. Voluntary for Double Organ Z719, bars 14–22

[72] Cox, *Organ Music in Restoration England*, i. 179–87.

Music of this sort provides us with a tantalizing glimpse of the body of organ music Purcell might have left had it been the practice for Restoration organists to write down their voluntaries. Roger North treated 'The Excellent Art of Voluntary' as an improvisation skill, and in particular wrote that 'sometimes the great performers upon organs will doe voluntary, to a prodigy of wonder, and beyond their owne skill to recover and set downe'.[73] Of the eminent organists of the Restoration period—William Child, Edward Lowe, Benjamin Rogers, John Hingeston, Christopher Gibbons, Matthew Locke, Albertus Bryan (Bryne), Vincenzo and Bartolomeo Albrici, Giovanni Battista Draghi, John Blow, Henry Purcell—only Blow left a sizeable body of organ music, and that, presumably, was mainly because he trained generations of boys in the Chapel Royal, and would therefore have needed a good deal of teaching material.

[73] Wilson, *Roger North*, 135–45, esp. 139.

IV
CHURCH MUSIC

WHEN Charles II returned to claim his kingdom in 1660 there had been little but simple psalm singing in England's churches for eighteen years. In 1642 Parliament had ordered 'such part of the Common Prayer and service as is performed by singing men, choristers, and organs' to be 'wholly forborn and omitted'; services were 'to be done in a reverent, humble and decent manner without singing or using the organs', and in 1644 an ordinance was issued for 'the speedy demolishing of all organs, images and all matters of superstitious monuments in all Cathedralls, and Collegiate or Parish-Churches and Chapels, throughout the Kingdom of England and the Dominion of Wales'.[1] Most organs were destroyed or dismantled, and church musicians were forced to fend for themselves. A generation grew up in the 1640s and 1650s that knew nothing of the great repertory of cathedral music. Samuel Pepys (b. 1633) wrote on 8 July 1660 that he had 'heard very good Musique' at Whitehall, 'the first time that I remember ever to have heard the Organs and singing-men in Surplices in my life'; on 4 November he wrote after a visit to Westminster Abbey that it was 'the first time that I ever heard the organs in a Cathedrall'.[2]

Only a few members of the pre-war Chapel Royal were still available for service in 1660, and took up their places once again, though the newcomers included William Child, Christopher Gibbons, and Edward Lowe, who had been organists at Windsor, Winchester, and Christ Church, Oxford.[3] Lowe felt it necessary to publish *A Short Direction for the Performance of the Cathedrall Service* in 1661 for 'The Information of such Persons, as are Ignorant of it, And shall be call'd to officiate in Cathedrall, or Collegiate Churches, where it hath formerly been in use'. It evidently met a need, for an enlarged edition taking account of the

[1] S. Eward, *No Fine but a Glass of Fine Wine: Cathedral Life at Gloucester in Stuart Times* (Wilton, 1985), 73; Scholes, *Puritans and Music*, 232.

[2] Pepys, *Diary*, i. 195, 283.

[3] Rimbault, *Old Cheque-Book*, 9–12, 128–9; Baldwin, *Chapel Royal*, 422–3.

1662 Prayer Book, *A Review of a Short Direction*, appeared in 1664 with a dedication to the Sub-Dean of the Chapel Royal.

There was much to do in the Chapel Royal, as elsewhere. The chapel, a small building roughly 75′ × 30′ in the north-east corner of old Whitehall, survived the Interregnum more or less intact, and seems to have been used during the Commonwealth.[4] But John Playford had to rescue the chapel's organ and music books before services could resume, and at first the results were not uniformly impressive: on 14 October 1660 Pepys heard 'an anthemne, ill sung, which made the King laugh'.[5] Part of the problem was that at first there were no trained boys. Pepys only mentioned 'singing-men' in his report of 8 July 1660, and Matthew Locke wrote that 'for above a Year after the Opening of His Majesties Chappell, the Orderers of the Musick there, were necessitated to supply the superiour Parts of their Musick with Cornets and Mens feigned Voices, there being not one Lad, for all that time, capable of Singing his Part readily'.[6] Luckily, the new Master of the Children, Captain Henry Cooke, excelled at his job. He recruited and trained a group of boys that included the future composers John Blow, Pelham Humfrey, Michael Wise, William Turner, Henry Hall, and Henry Purcell.

In the Chapel Royal, as in other 'quires and places where they sing', the main function of the choir was to participate daily in Morning and Evening Prayer. In addition, on 'Litany Days' (Sunday, Wednesday, and Friday) Morning Prayer concluded with the Litany, while the 'Second Service', Holy Communion, was added on Sundays and holy days.[7] In the *Short Direction* Lowe provided plainsong melodies, either unaccompanied or in simple four-part harmonizations, for 'the Ordinary and Extraordinary parts' of these services, 'both for the Priest, and whole Quire'. But this was a stopgap. He wrote that 'The Tunes in foure parts' were to serve 'only so long till the Quires are more learnedly Musicall, and thereby a greater variety used.'

The 'greater variety' took various forms. The plainsong tones for the psalms were gradually replaced by a new repertory of

[4] Holman, *Four and Twenty Fiddlers*, 389–93.

[5] A. Freeman, 'Organs Built for the Royal Palace at Whitehall', *MT* 52 (1911), 720–1; Pepys, *Diary*, i. 265–6.

[6] M. Locke, *The Present Practice of Musick Vindicated* (London, 1673; repr. 1974), 19.

[7] For the liturgy, see P. le Huray, *Music and the Reformation in England, 1549–1660* (London, 1967), 157–63; C. Dearnley, *English Church Music 1650–1750* (London, 1970), 96–102.

composed four-part Anglican chants, including some attributed on more or less tenuous grounds to Henry Purcell.[8] Similarly, a new repertory of composed settings soon developed for the corpus of texts called the service. The full Restoration service consisted of the canticles (the psalms and other texts prescribed for daily use, as opposed to those that vary from day to day) in Morning and Evening Prayer, and certain texts in the Ordinary of Holy Communion. In the Morning Service the texts set were the Te Deum and Jubilate Deo, with the Benedictus and Benedicite as alternatives. In the Evening Service they were the Magnificat and Nunc dimittis, with the Cantate Domino and the Deus misereatur as alternatives. The texts in the Communion Service were the Kyrie, the Creed, and, occasionally, the Sanctus and the Gloria. Most composers confined themselves to the Te Deum and Jubilate, the Magnificat and Nunc dimittis, or the Kyrie and Creed, but Purcell's Service in B flat major Z230 includes the alternative canticles as well.

We do not have a source that gives us an order of service for the Restoration Chapel Royal, but the next best thing is a passage in the second, enlarged edition of *The Divine Services and Anthems Usually Sung in his Majesties Chappell, and in all Cathedrals and Collegiate Choires in England and Ireland*, published in 1664 by James Clifford, at the time a minor canon of St Paul's Cathedral.[9] His directions for 'that part of the Divine Service perform'd with the organ at S. Paul's Cathedrall on Sundays, etc.' establishes the place of the anthems in the service, and also provides scarce information about the role of the organ in the liturgy:

The First Service in the Morning.

After the Psalms a Voluntary upon the organ alone. After the Ist Lesson is sung Te Deum Laudamus . . . After the 2nd Lesson, Benedictus . . . or Jubilate Deo . . . After the 3rd Collect . . . is sung the Ist Anthem. After that the Litany . . . After the Blessing . . . a Voluntary alone upon the organ.

The Second or Communion Service.

After every commandment, the Prayer, 'Lord have mercy upon us,' etc. . . . After the Epistle, this heavenly ejaculation, 'Glory be to Thee, O Lord'. After the Holy Gospel, the Nicene Creed . . . After the sermon, the last Anthem.

[8] Zimmerman, *Purcell: Catalogue*, 75–8, 408–10.

[9] For Clifford, see Shay, 'Henry Purcell', 51–4; for a list of the anthem texts, see ibid. 240–7.

At Evening Service.
After the Psalms, a Voluntary alone by the organ. After the Ist Lesson is sung the Magnificat . . . After the 2nd Lesson, the Nunc Dimittis . . . or Deus Misereatur . . . After the 3rd Collect . . . is sung the first Anthem. After the sermon is sung the last Anthem.[10]

In addition, the organ was also used at the beginning and end of services, as it is today. The Revd John Newte preached a sermon at Tiverton on 13 September 1696 defending the use of the organ in the liturgy, in which he referred to 'some taking Lesson or decent Flourish or other by itself, which goes by the Name of Voluntaries'.[11] 'This sort of Musick' was heard 'just before the Service begins . . . to engage the Congregation to a serious Thoughtfulness', 'at the end of the Psalms, before the Lessons be read, to strike a reverential Awe upon our Spirits, and to melt us into a fit Temper to receive the best Impressions from the Word of God', and 'at the end of the whole, to take off some little whispering Disturbances, through the Levity of some People, and to drown that ungrateful rushing Murmur and Noise which the stirring of so many People together, at that time of going out of the Church, must occasion'.

The service was the type of church music least affected by the changes of musical style that followed the Restoration. Indeed, it is often difficult to be sure whether examples by the older generation were written before 1642 or after 1660. As Thomas Tudway (another of Cooke's choirboys) put it:

The Old Masters of Music viz: Dr Child, Dr Gibbons, Mr Law [actually Lowe], &c Organists to his Majesty, hardly knew how, to comport themselves, w(i)th these new fangl'd ways, but proceeded in their Compositions, according to the old Style, & therfore, there are only some services, & full Anthems of theirs to be found.[12]

But the old masters did introduce some innovations. There was no longer much of a distinction between the simple short service in full style (purely choral), and the large-scale and elaborate great service in verse style (solos and ensembles alternating with full sections). Most Restoration services have verse passages, and yet the prevailing style is simple and functional, partly, no doubt,

[10] J. S. Bumpus, *A History of English Cathedral Music 1549–1889* (London, 1889), i. 117–18.

[11] Cox, *Organ Music in Restoration England*, i. 6–7, 10–11.

[12] C. Hogwood, 'Thomas Tudway's History of Music', in Hogwood and Luckett (eds.), *Music in Eighteenth-Century England*, 25–6.

because Charles II was notoriously averse to counterpoint. Instead, interest is maintained by constantly varying the combinations of solo voices in the verse sections, and by changing from the prevailing sober duple time to a jaunty triple time, the king's 'step tripla'. Some services, such as the Evening Service in G minor Z231 attributed to Purcell, are in triple time throughout.[13] Also, by the 1670s, when the new generation was contributing to the repertory, elements of the expressive harmonic style of the period—false relations and other unprepared dissonances, chromaticism, and unpredictable, rapid modulations—were even filtering into conservative genres such as the service.

There is counterpoint in Restoration services, but it tends to be concentrated in the doxologies (the 'Glory be to the Father' sections) of the canticles, and it is mostly of a specialized type. Ian Spink has suggested that William Child's 'Sharp Service' was the origin of the Restoration fashion for canons in services, but the immediate model for the nine in Purcell's B flat Service seems to have been Blow's G major Service.[14] Purcell certainly knew the work well. He copied the organ part of the Sanctus and Gloria into a Chapel Royal organ-book, and he included the doxology of the Jubilate, a canon 'four in one', as one of his counterpoint examples in the 1694 edition of Playford's *Introduction to the Skill of Musick*.[15] He added: 'this very Instance is enough to recommend him for one of the Greatest Masters in the World'. Posterity agreed with him: it can be seen carved in stone at the base of Blow's monument in Westminster Abbey.

Purcell was mostly content to write the same sort of canons as Blow: three in one, four in one, and four in two (respectively, a simple three-part canon, a four-part canon, and a double canon), though the pupil outdid the master by including inversion canons two in one, three in one, four in one, and four in two. He avoided setting the doxology of the Jubilate as a canon, presumably as a mark of respect to his teacher, though he modelled his doxology canon in the Magnificat on the one at the same point in Blow's service. Both are three-in-one canons over a free bass, and have

[13] I am grateful to Bruce Wood for pointing out to me that Z231 survives only in late sources, and is of questionable authenticity.

[14] I. Spink, 'Church Music II: From 1660', in Spink (ed.), *Seventeenth Century*, 104; the Child and part of the Blow are in W. Boyce (ed.), *Cathedral Music* (2nd edn., London, 1788), i. 33–68; iii. 252–88.

[15] H. W. Shaw, 'A Cambridge Manuscript from the English Chapel Royal', *ML* 42 (1961), 263–7; Squire, 'Purcell as Theorist', 566–7.

similar themes. Their progressions are also similar at the beginning, though later Purcell creates much more harmonic drive than Blow by constantly modulating as the canon develops (Ex. 4.1).

Ex. 4.1. 'Glory be to the Father' from the Magnificat in B flat major Z230/7g, bars 130–46

In his youth, Purcell seems to have needed strict counterpoint to keep his angular and unpredictable melodic and harmonic style under control; the tension thus created between unconventional wine and conventional bottles is often powerfully expressive. As he got older his interest in formal contrapuntal devices lessened as he came to terms with the Italian style, with its predictable phrase patterns, smooth melodic lines, and logical, directional harmonies. Nevertheless, he never lost interest in canon—notable late examples are the Chaconne for two recorders and continuo ('Two in one upon a Ground') in *Dioclesian* Z627/16 and the Dance for the Followers of Night in *The Fairy Queen* Z629/15 (four in two)—and he placed considerable emphasis on the genre in the 1694 *Introduction to the Skill of Musick*, calling it the 'noblest sort of Fugeing'.

Head-motifs (ideas placed at the beginning of several movements of an extended liturgical work to establish an audible relationship between them) are usually associated with Renaissance masses, but they are often found in Tudor and early Stuart services, and the tradition continued after the Restoration. Some of the movements of Child's Service in E minor are linked in this way, as are those of Benjamin Rogers's Service in D major.[16] The most obvious links in Purcell's B flat Service are between the opening of the Benedictus and the Nunc dimittis, and between the opening of the Te Deum and the Cantate Domino. Similarly, there are obvious links between the opening of the Te Deum, the Kyrie, the Magnificat, and the Nunc dimittis in Blow's G major Service.

But Purcell goes a stage further by linking some movements of the B flat Service with his anthem 'O God, thou art my God' Z35; a primary source of the latter, in GB-Cfm, Mu. MS 117, has the note 'To Mr Purcell's B-mi Service'.[17] The anthem mostly shares ideas with the canticles of the evening portion of the service, which suggests that it was written to be performed with them. The relationship between the two works goes far beyond the use of head-motifs, and is similar to that between a parody mass and its model, though it is not clear which came first; Zimmerman suggests they were composed together.

The other significant early liturgical work is a setting of part of

[16] Boyce, *Cathedral Music*, i. 145–69, 170–93.

[17] F. B. Zimmerman, 'Purcell's "Service Anthem" "O God, Thou art my God", and the B flat major Service', *MQ* 50 (1964), 207–14; see also Shay, 'Henry Purcell', 160–5.

the Burial Service. In its complete form it consists of the sentences 'Man that is born of a woman', 'In the midst of life', and 'Thou know'st Lord', but the early autograph in GB-Lbl, Add. MS 30931 lacks the first, while the later autograph copy of a revised version in GB-Cfm, Mu. MS 88 lacks the third—which, however, exists in secondary sources. Purcell's revisions, apparently carried out in the late 1670s, range from changes of detail to radical recomposition, and offer fascinating glimpses of the composer at work, improving the part-writing, as well as clarifying and modernizing the harmony.[18] It has been assumed that he also added 'Man that is born of a woman' at the same time. But all the sentences would have been needed for a liturgical performance, and it is therefore more likely that an early autograph of the first verse has been lost. Zimmerman confused the issue when he gave the first and second versions of 'In the midst of life' and 'Thou know'st Lord' the numbers Z17A and B and Z58A and B, but also allocated Z27 to the second version of the complete work.

The idiom of Purcell's early burial sentences—open textures, wild dissonance, jagged vocal lines—is close to that of the early sacred part-songs, discussed in Ch. 2. The Prayer Book requires their texts to be sung 'When they are come to the Grave, while the Corpse is made ready to be laid into the earth', but they have solos that require an organ accompaniment, so Bruce Wood has recently suggested that they were written for a burial inside Westminster Abbey, such as the funeral of Pelham Humfrey (1674), or even that of Henry Cooke (1672), though the funeral of Christopher Gibbons (1676) is perhaps more likely, for they are remarkably assured even in their first versions.[19] 'Remember not, Lord, our offences' Z50, a setting of part of the 'Order for the Visitation of the Sick' from the Prayer Book, has sometimes been associated with them, though it is in a different key (A minor rather than C minor) and has a different scoring (SSATB full instead of SATB verse and full). With its mixture of block chords and simple counterpoint, it is also closer to the note-against-note style of earlier settings of the burial sentences.

Purcell seems to have written all his liturgical music early in his

[18] R. Manning, 'Revisions and Reworkings in Purcell's Anthems', *Soundings*, 9 (1982), 29–37; R. Ford, 'Purcell as his own Editor: The Funeral Sentences', *Journal of Musicological Research*, 7 (1986), 47–67; Shay, 'Henry Purcell', 134–54.

[19] B. Wood, 'The First Performance of Purcell's Funeral Music for Queen Mary', in Burden (ed.), *Performing the Music of Purcell*.

career, apart from a couple of pieces for special occasions: the D major Te Deum and Jubilate Z232, and the late setting of 'Thou know'st, Lord, the secrets of our hearts' (which will be discussed later). Purcell's B flat Service cannot be dated exactly, but it was in existence in 1682, when 30*s*. was spent at Westminster Abbey for having 'm^r^ Purcells Service & Anthems' copied; an early copy survives in the Abbey library.[20]

It is understandable that Purcell might have lost interest in such a restricted and conventional idiom as the service after the early 1680s, but it is puzzling that he wrote so few full anthems, given his interest in formal counterpoint. The only surviving Purcell anthems without verse passages are 'Hear my prayer, O Lord' Z15 (SSAATTBB), 'O God, the king of glory' Z34 (SATB), 'O God, they that love thy name' ZD4 (?SATB), 'Thy righteousness, O God, is very high' Z59 (?SATB), and the recently discovered 'I was glad when they said unto me' (SSATB).[21] The first two seem to be fragments of larger works, which could have had verse passages, while the next two only survive incomplete in late sources, and may be spurious. Thus, 'I was glad' may be Purcell's only true full anthem, and it is significant that it was written for a special occasion, James II's coronation in 1685. The same pattern can be observed in Blow's output: he seems not to have written any full anthems between 1682 and 1697, apart from pieces for the 1685 and 1689 coronations.[22] The implication is that Charles II's musical taste prompted his composers to concentrate almost entirely on the verse anthem. In turn, it suggests that Purcell wrote less for Westminster Abbey than has been thought. Blow only returned to the full anthem in the late 1690s, once he had resumed his post there after Purcell's death.

The verse anthem was invented in Elizabeth's reign. It began as an offshoot of the secular repertory of 'consort songs' for boy soloists and viols, mostly composed by Chapel Royal composers for choirboy plays. The earliest verse anthems are virtually sacred consort songs, with short full sections taking up the words and music of the solos. In James I's reign composers began to use var-

[20] Shay, 'Henry Purcell', 59.

[21] Purcell, *I Was Glad when They Said unto Me*, ed. B. Wood (London, 1977); Wood, 'Two Purcell Discoveries—2: A Coronation Anthem Lost and Found', *MT* 118 (1977), 466–8.

[22] Spink, 'Church Music II', 112.

ied groups of soloists, adults as well as boys, and developed the full sections, often giving them their own material. By then verse anthems had been taken up by other collegiate foundations, where viols were usually not available. In some places they were accompanied by wind consorts, but they were also commonly arranged for organ. In time, anthems came to be written specifically with organ, though the keyboard parts were still written out as if they were reductions of ensemble parts. Continuo playing became common among English organists in the 1650s, and so the practice of writing out organ parts declined after the Restoration, though there are still some examples in early Purcell. It became more common again in the 1690s when strings were no longer available for the Chapel Royal, and the organ was used to play a reduction of their parts in symphony anthems.

The change in the type of accompaniment brought changes to the solo vocal parts. In pre-war anthems they were essentially strands in a contrapuntal texture, whether they were accompanied by a consort of instruments or the organ. Substituting a continuo line encouraged the use of declamatory writing and florid ornamentation, as well as abrupt changes of time, pace, and mood—which were most easily accompanied from a simple bass part. Much has been made of the supposed influence of French and Italian church music on the Restoration anthem, but it must be remembered that English song composers had made similar changes in the reign of James I. Significantly, the earliest datable continuo anthem, Walter Porter's 'O praise the Lord', was published in a secular collection, his *Madrigales and Ayres* (London, 1632).[23] Henry and William Lawes, the leading song composers of their generation, also wrote some early examples.[24]

A number of pieces have been proposed as Purcell's first continuo anthem. Peter Dennison's candidate was 'Turn thou us, O good Lord' Z62 (ATB verse, SATB full) 'on the grounds of its highly derivative style and uneven quality', but the solo sections for a solo tenor use a fairly sophisticated and elaborate form of declamatory writing, and the ATB verse passage has some effective chromaticism.[25] A better candidate is 'Lord, who can tell

[23] P. le Huray (ed.), *The Treasury of English Church Music*, ii: *1545–1650* (London, 1965), 232–47.

[24] See e.g. William Lawes's 'The Lord is my light', in Boyce (ed.), *Cathedral Music*, ii. 219–25.

[25] P. Dennison, 'Purcell: The Early Church Music', in Sternfeld *et al.* (eds.), *Essays on Opera and English Music*, 54.

how oft he offendeth?' Z26 (TTB verse, SATB full), a setting of Ps. 19: 12–14. It is much simpler in style and form than any other Purcell anthem, and it occurs in a fragmentary set of Chapel Royal part-books probably copied not later than 1678, and possibly around 1676.[26] It consists of a single extended verse section followed by a concluding chorus. The duple-time solo passages are plain and simple, with only a few declamatory mannerisms and virtually no affective harmony. They could almost come from an anthem by Henry Lawes or one of his contemporaries, and another archaic feature is the presence of several short passages for continuo alone, framing the solos; they mostly anticipate or repeat the harmonies of the neighbouring vocal phrases, and may be the remnants of a written-out organ part.

'O Lord our governor' Z39 (SSSBB verse, SATB full), a setting of Ps. 8, is also very early. It was copied by Stephen Bing (d. November 1681) into GB-Y, M.1.S, the so-called 'Gostling Part-Books', and its style suggests that it was written not long after Z26.[27] It starts with a written-out organ prelude in imitative style, but once the soloist enters for the first verse the accompaniment becomes a figured bass. Composers of the older generation, such as Christopher Gibbons and Edward Lowe, sometimes combined continuo parts with written-out entries in this way. The harmonic style is still fairly simple, though there are one or two affective mannerisms (Ex. 4.2). The example contains an unprepared dominant seventh, as well as a characteristic Restoration cadence where the voice anticipates the minor third of a tonic chord, only to settle unexpectedly on the *tierce de Picardie*. The solo sections also have more complex melodic writing than Z26, with a few bursts of florid ornamentation.

Purcell was aware that a work of more than 200 bars would be monotonous if he began and ended each main section in the home key (C minor) as tradition demanded, so he placed one in G minor and modulated a good deal within the sections. For the same reason, he tried to vary the rhythmic movement and the pace—there are eight changes to and from triple time—though it must be said that the work is over-long and rather incoherent, and there are places where the harmonic direction is in doubt.

[26] H. W. Shaw, 'A Contemporary Source of English Music of the Purcellian Period', *AcM* 31 (1959), 38–44.

[27] Id., *The Bing–Gostling Part-Books at York Minster* (Croydon, 1986); P. Willetts, 'Stephen Bing: A Forgotten Violist', *Chelys*, 18 (1989), 3–17; Shay, 'Henry Purcell', 64–8.

Ex. 4.2. 'O Lord our governor' Z39, bars 11–21

Nevertheless, there are some charming moments, as when three trebles warble in close harmony 'Out of the mouths of very babes and sucklings hast thou ordained strength', or when two basses gravely sing a semi-canonic setting of 'O Lord our governor, how excellent is thy name in all the world'.

The next stage of Purcell's development as an anthem composer is represented by GB-Cfm, Mu. MS 88, the earliest of his three large autograph score-books. He devoted the back of the volume to full anthems or verse anthems with organ, including a number by pre-Civil War composers (see Ch. 1). Towards the end he began to include items of his own, starting with the verse anthem 'Save me, O God' Z51 (INV, no. 17). They may well be the first he thought worth preserving; they are certainly an advance on the ones just discussed, and in general Mu. MS 88 enables us to look at the Restoration anthem through the young Purcell's eyes.

The second piece of his own in the sequence (INV, no. 20) is 'Blessed is he whose unrighteousness is forgiven' Z8 (SSATTB verse, SATB full), a setting of verses from Ps. 32. It is a fine piece, with some expressive six-part verse-writing in declamatory style at the beginning. But it is still rather rambling and incoherent: it runs to nearly 200 bars, and is divided into nine separate sections. 'Bow down thine ear, O Lord' Z11 (SATB verse and full; MS 88 INV, no. 23), a setting of verses from Ps. 86, is similar in style, though more tautly constructed. It uses a common double-barrelled pattern: after an introductory quartet each of the two halves progresses from a solo (tenor in the first, bass in the second) to a quartet, and then to a full passage. The works starts, like Z39, with the vestige of a written-out organ part, but the four-part verse that follows belongs to a different, more modern world. It is laid out as a series of orthodox imitative entries, setting the subject to 'Bow down thine ear, O Lord, and hear me', and the countersubject to 'for I am poor and in misery'. But the music is anything but orthodox. The melodic shapes are angular and syllabic, with the countersubject in declamatory patterns of quavers and semiquavers, and there is constant modulation (nine keys are touched on in eighteen bars), driven foward by chromaticism and a remarkable series of false relations created by upward-resolving appoggiaturas on the word 'poor' (Ex. 4.3).

I have already suggested that music of this sort owes little to Italian music. Some printed collections of Italian motets were imported into seventeenth-century England, and were used by English composers to make their own manuscript collections; Locke's own collection, copied in the Netherlands in 1648, survives in GB-Lbl, Add. MS 31437.[28] They were doubtless used as formal models when Locke and his contemporaries wrote their own Latin motets, but they do not contain much of the flamboyant harmonic writing and angular melodic writing that abounds in Restoration church music. In fact, all the ingredients of the style—declamatory vocal lines, florid ornamentation, expressive and dissonant harmony, rapid modulation—had long been present in English song. Furthermore, the composer most responsible for introducing it into Restoration church music was the Catholic Matthew Locke, who was only peripherally involved with the Chapel Royal, and wrote more secular music than sacred. He was a distinguished composer of theatre music and songs, and was steeped in the declamatory tradition.

Locke's contribution to the idiom of Restoration church music

[28] Harding, *Thematic Catalogue*, 143; for English copies of Italian music, see D. Pinto, 'The Music of the Hattons', *RMARC* 23 (1990), 79–108; J. P. Wainwright, 'George Jeffries' Copies of Italian Music', ibid. 109–24.

Ex. 4.3. 'Bow down thine ear, O Lord' Z11, bars 1–18

Verse

Bow down thine ear O

5

Lord and hear me, for I am poor and in mi - se-ry,

Bow down thine ear O Lord and hear me,

Bow

8

for I am poor and in mi - se-ry, am

for I am poor_ and in mi - se-ry, bow

down thine ear O Lord and hear me, for I am

Bow down thine ear O

Ex. 4.3. *cont.*

11
[tr]
poor _ and in mi - se - ry, bow
down thine ear O Lord and hear me,
[tr]
poor _ and _ in mi - - se - ry, in mi - - se -
Lord and hear me, for I am
13
down thine ear O Lord and hear me, hear ___ me, for I am
for I am poor _ and in mi - se-ry,
-ry, bow down thine ear O Lord and hear ___ me, and
poor _ and in mi - se-ry, for I am poor ___
16
[tr]
poor, _ am poor and _ in mi-se-ry, I am poor ___ and in mi - se-ry.
for I am poor ___ and _ in mi-se-ry, and in mi - se-ry.
[tr]
hear ______ me, for I am poor _ and in mi - se-ry.
and in mi - se-ry, I am poor and in mi - se-ry.

can be assessed by comparing three examples in Mu. MS 88 of the 'full with verse' anthem, a special type sometimes regarded as a form of full anthem. Purcell was particularly attracted to it in the late 1670s, and copied examples by Locke and Blow in the collection, as well as six of his own. It is essentially contrapuntal and full-voiced. The full sections outnumber the verse passages—common patterns are Full–Verse–Full or Full–Verse–Full–Verse–Full—and the soloists nearly always sing in ensemble, often with groups of high and low voices in dialogue, as in many services. The organ is not always specifically indicated, and when it is it acts as a basso seguente, doubling the lowest voice rather than providing an independent bass.

The three works, Blow's 'O God, wherefore art thou absent from us' (SSATB verse and full; INV, no. 3), Locke's 'Turn thy face from my sins' (SSATB verse and full; INV, no. 7), and Purcell's 'O God, thou hast cast us out' Z36 (SSAATB verse and full; INV, no. 27), are remarkably similar in outline.[29] All three set penitential texts—the Blow Ps. 74: 1–3, the Locke Ps. 51: 9–13, and the Purcell Ps. 60: 1–2, 11–12—and have verses enclosed by full sections, the Locke and the Purcell in the simple Full–Verse–Full pattern (though the editor of the Locke suggests, surely wrongly, that the first section should be verse), the Blow in the Full–Verse–Full–Verse–Full pattern with the opening imitative point returning at the end.

The Blow and the Purcell are much more old-fashioned than the Locke, though they were probably written some time after it. They use a straightfoward contrapuntal style in the full sections. The lines consist mainly of simple patterns of minims and crotchets, the rate of chord change is relatively slow, dissonance is mostly produced by prepared suspensions, and the word-painting arises in the traditional way out of the shape of the phrase: a falling diminished fourth at the words 'absent from us' in the first section of 'O God, wherefore art thou absent from us'; a mountain-shaped arpeggiated phrase for 'Mount Sion' in the last section; striding broken chords in 'O God, thou hast cast us out' for 'thou hast cast us out and scatter'd us'; and, in the last section, majestic falling scales for 'he shall tread down our enemies'. Purcell presumably had Blow's anthem in mind when he began to compose, for they both start in a similar way: the words 'O God'

[29] J. Blow, *O God, Wherefore Art Thou Absent from Us* (Chichester, 1990); M. Locke, *Anthems and Motets*, ed. P. le Huray (MB 38; London, 1976), no. 14.

are set to semibreves, followed by a rest (into which the first note of the next entry is inserted), and then by the rest of the sentence in a crotchet rhythm (Ex. 4.4).

By contrast, Locke's anthem starts with an imitative point in a simple pattern of minims and crotchets, but it continually twists and turns from key to key, illustrating the words 'Turn thy face' by means of the harmony rather than the melodic shapes. Rapid,

Ex. 4.4*a*. John Blow, 'O God, wherefore art thou absent from us', bars 1–23

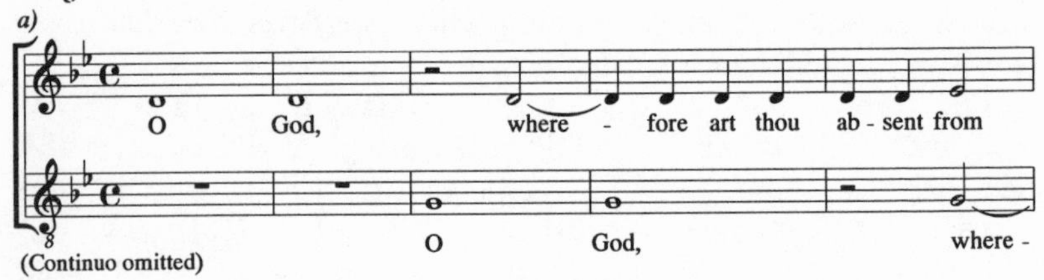

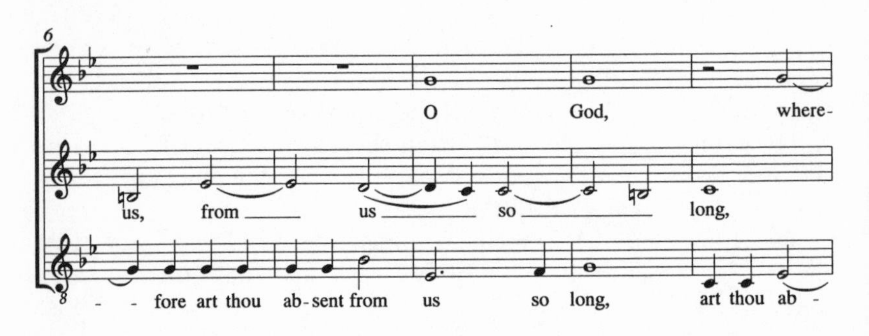

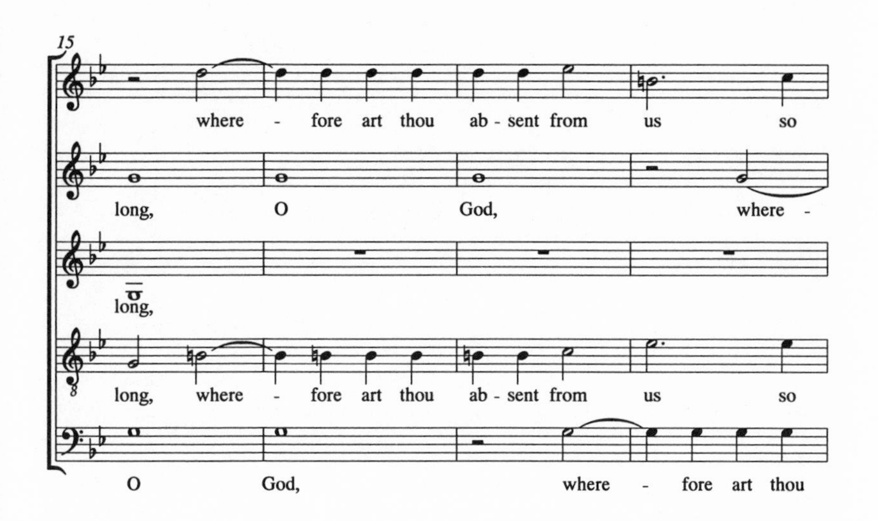

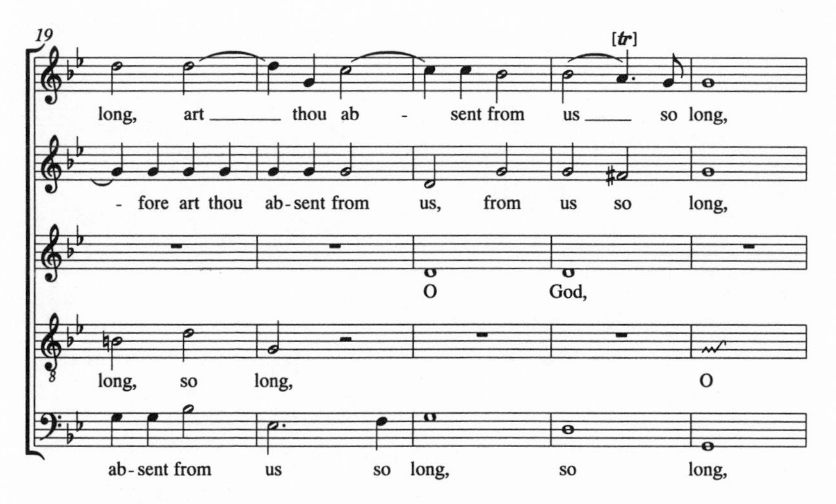

Ex. 4.4*b*. 'O God, thou hast cast us out' Z36, bars 1–17

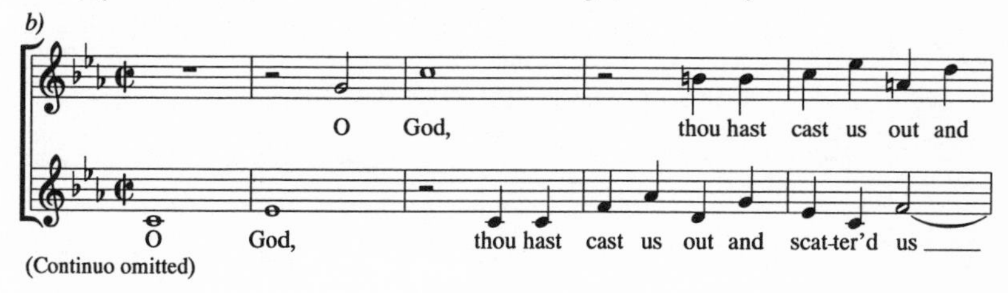

Ex. 4.4 *cont.*

6
scat - ter'd us a - broad,
O God,
a - broad,
O God, thou hast
O God,
10
O God,
thou hast cast us out and scat-ter'd us a - broad, O God,
O God, thou hast
cast us out and scat-ter'd us a - broad,
thou hast cast us out and scat-ter'd, scat - - - - ter'd
14
thou hast cast us out and scat - ter'd us a - broad,
thou hast cast us out and scat - ter'd us
cast us out and scat - ter'd us a - broad,
us a - broad,

nervous modulation of this sort is a hallmark of his style, and he combines it in the verse section with a range of unorthodox dissonances, including unprepared fourths and dominant sevenths, and several examples of what might be called the 'English sixth' (a 6♭–3♯ chord), a favourite with Restoration composers. At the climax of the passage there is a delicious complex of false relations and seventh chords illustrating the word 'sinners' (Ex. 4.5). The colourful 7–5♯–3 chord on the first beat of bar 56, effectively a dominant chord combined with a mediant, was much used in France at the time, but is rare in English music. The melodic shapes are no less remarkable. They have the typical angular, asymmetrical patterns of declamatory song, and yet they are subject to the discipline of traditional imitative counterpoint.

Much of Locke's music is concerned with reconciling the seemingly irreconcilable in this way, and Blow and Purcell learnt from him that it was possible to achieve a fruitful synthesis between formal counterpoint and expressive, soloistic vocal writing—between the English versions of the *stile antico* and the *stile moderno*. Much of Blow's music of the 1680s is extremely angular and unpredictable, and his supposed lapses (including the 'English sixth') so offended Charles Burney that he went to some trouble to compile a collection of examples of 'Dr Blow's Crudities', some of which he published in his *History*.[30]

Purcell, too, went through an experimental phase around 1680, though he only rarely attracted criticism from Burney. (It would be nice to think that the censorious music historian responded directly to Purcell's discriminating ear, but one suspects that it had more to do with his conception of 'our British Orpheus, or rather our musical Shakespeare' as the only Restoration composer whose music had transcended 'the erosions of time and vicissitudes of taste'.[31]) There are some remarkable things in 'Lord, how long wilt thou be angry?' Z25 (ATB verse, SSATB full; Mu. MS 88 INV, no. 31), a setting of verses from Ps. 79, but the finest and most daring of the full with verse anthems are the two for an unusually large number of voices, the eight-part 'O Lord God of hosts' Z37 (SSAATB verse, SSAATTBB full; Mu. MS 88 INV, no. 29), and the ten-part (actually only in eight real parts) 'Blow

[30] Burney, *General History*, ii. 351–5.

[31] R. Luckett, '"Or Rather Our Musical Shakespeare": Charles Burney's Purcell', in Hogwood and Luckett (eds.), *Music in Eighteenth-Century England*, 72–3.

Ex. 4.5. Matthew Locke, 'Turn thy face from my sins', bars 49–58

Ex. 4.5. *cont.*

up the trumpet in Sion' Z10 (SSSATTB verse, SSSAAATTBB full).

Locke is certainly the inspiration for 'Blow up the trumpet in Sion'. The work is not in Mu. MS 88, but was apparently composed by the autumn of 1677, for it comes in part-books at Westminster Abbey before items attributed to 'Mr Blow'.[32] The text, from Joel 2: 15–17, is an urgent call for a sinful people to turn to God in the face of danger from the heathen, and may well have been chosen for a special occasion; it would have been appropriate in the fevered atmosphere of the spring and summer of 1677, when a string of French military and naval victories caused great alarm in England. Purcell's music certainly has a sense of occasion, the more so since it seems to allude to Locke's anthem 'Be thou exalted Lord', performed in the Chapel Royal on 14 August 1666 to celebrate a naval victory over the Dutch.[33]

Purcell does not use Locke's magnificent polychoral writing, with five separate ensembles, two instrumental and three vocal; Purcell and Blow always preferred to lay out their eight-part choral writing in a single, sonorous group, even though the layout of sets of cathedral part-books allow for the division into separate Decani and Cantoris choirs. But the seven-part verse passages, with their declamatory writing tossed between ever-changing

[32] Information kindly supplied by Bruce Wood; see also Shaw, *Bing–Gostling Part-Books*, 108–11.

[33] Locke, *Anthems and Motets*, no. 7; see also Holman, *Four and Twenty Fiddlers*, 403–4.

groups of solo voices, are strikingly similar to Locke's verse passage—also seven-part—starting at the words 'Thou hast given him his heart's desire', and Purcell follows Locke in largely abandoning formal counterpoint in favour of what can best be described as kaleidoscopic choral declamation. But, as so often in Locke, it is Purcell's harmony that remains in the memory: the dramatic change from the opening C major fanfares to an E flat chord at 'sanctify a fast', the daring chromatic writing at 'Let them weep between the porch and the altar', and the final cadence at 'spare thy people, O Lord', with its eloquent false relations. Given its difficulty, it is not surprising that 'Blow up the trumpet in Sion' did not circulate in the wider cathedral repertory, and it is rarely heard today. Nevertheless, it is one of Purcell's finest anthems.

'O Lord God of hosts', a setting of verses from Ps. 80, takes as its starting-point Blow's eight-part 'God is our hope and strength' (Mu. MS 88, INV, no. 2). Purcell copied Blow's scoring right down to the precise layout of the first verse passage, SSA in dialogue with ATB, and also used the same unusual key, A major—which would have produced some piquant effects whenever the music moved outside the limited cycle of keys encompassed by mean-tone organ tuning. However, Purcell's anthem far outshines its model: sonorous, vigorous counterpoint for the eight-part choir makes an admirable foil to a series of daring progressions in the verse passages. In the excerpt given in example 4.6 one can only marvel at the acute and inventive harmonic imagination that could produce music so seemingly logical and inevitable from such an extraordinary collection of chords. Purcell probably wrote this powerful piece in the late 1670s, for it appears in a list of pieces added to the Chapel Royal books between 1677 and Christmas Day 1680.[34] By then he had little left to learn from Blow or Locke.

We have no more full with verse anthems by Purcell after 'O Lord God of hosts', and in fact the period around 1680 was something of a watershed for the Chapel Royal repertory. Blow and Purcell seem virtually to have abandoned the various types with organ accompaniment in favour of symphony anthems. There are only a handful of continuo anthems by Purcell that cannot be dated either before then or after 1688, when the genre

[34] Ashbee, *RECM* i. 193–4.

Ex. 4.6. 'O Lord God of hosts' Z37, bars 38–47

was revived in a new form to cope with the absence of string-players in the Chapel Royal. Some of the continuo anthems ascribed to this period in Zimmerman's catalogue, such as 'I will sing unto the Lord' Z22 ('before 1683'), and 'Let God arise' Z23 ('after 1683'), are in Bing's hand in the Gostling Part-Books and were therefore written before November 1681. Others, such as 'O consider my adversity' Z32, sound like early pieces but only survive in later sources, while the popular 'Thy word is a lantern unto my feet' Z61 ('not later than 1687') circulated in the cathedral repertory as a continuo anthem but looks suspiciously as if it originally had string parts; the same thing seems to have happened to 'Thy way, O God, is holy' Z60 and 'O give thanks unto the Lord' Z33; see below.

The symphony anthem is the most glamorous type of Restoration church music, and has received the most attention from modern writers. It owed its existence to the personal taste of Charles II, 'a brisk, & Airy Prince, comeing to the Crown in the Flow'r, & vigour of his Age', who, in the famous words of Thomas Tudway, was soon 'tyr'd w(i)th the Grave & Solemn way, And Order'd the Composers of his Chappell, to add

Symphonys &c w(i)th Instruments to their Anthems'.[35] But it was always something of an exotic plant. It was rarely cultivated outside Whitehall, and it hardly outlived the monarch who created it. Also, Tudway added that symphony anthems were only performed when Charles II 'came himself to the Chapell, w(hi)ch was only upon Sundays in the morning, on the great festivals, & days of Offerings'.

The new form seems to have come into being as a result of the change from the old type of verse anthem to the continuo anthem. Once the solo passages began to be accompanied by continuo the way was open for the consort of instruments to be given a new role, alternating with the voices instead of accompanying them (though they continued to support the choir in the full sections), and this became the dominant pattern in the Restoration period. A similar change had been made in secular music in the 1630s. In Jacobean consort songs the viols accompany the voices, but in Walter Porter's *Madrigales and Ayres* the violins mostly alternate with them, playing symphonies and ritornellos. It is easy to imagine that the same thing happened in the Chapel Royal at the same time, though the first hard evidence is a group of anthem-like pieces by Henry Lawes in his autograph score, GB-Lbl, Add. MS 31434.

Several of them have short instrumental passages in two parts, treble and bass, marked 'symphony'. They must originally have had fuller string-writing, for the words of two of the anthems are in a printed pamphlet, *Select Psalmes of a New Translation, to be Sung in Verse and Chorus of five Parts, with Symphonies of Violins, organ, and other Instruments, Novemb(er) 22 1655. Composed by Henry Lawes, Servant to his Late Majesty*.[36] In one anthem the instrumental passages come from a five-part air by William Lawes. These pieces were presumably written for private devotions during the Commonwealth; the surviving copy of the pamphlet in US-SM belonged to the Bridgewater family, Lawes's patrons.

After the Restoration groups of string-players began to play in the Chapel Royal on a regular basis. At first there were only three of them, two violins and bass viol (which explains why a number of early symphony anthems only have three-part string-writing),

[35] Hogwood, 'Thomas Tudway's History of Music', 25.

[36] D. Pinto, 'The Fantasy Manner', *Chelys*, 10 (1981), 19–21, and C. Bartlett's notes to the Consort of Musicke's recording, *Henry Lawes: Sitting by the Streams, Psalmes, Ayres, and Dialogues*, Hyperion A66135 (1984).

but there were certainly four by 1667; regular rosters of five were established in 1670, and lasted (apart from a period in 1672 when six were used) until the end of Charles II's reign.[37] Theorbos are known to have been used in the Chapel, and several members of the Twenty-four Violins played the instrument, so the most likely combinations are string quartet and theorbo in anthems with four-part strings, or two violins, bass, and two theorbos in anthems with three-part strings. In addition, there are payments to two bass viol-players for playing in the Chapel in the 1660s and 1670s, but they are always listed separately from the other string-players, and they probably had some discrete function, such as doubling the bass of the Decani and Cantoris (called Sub-decani in the Chapel) sections of the choir. It seems that the violins had their own gallery in the chapel at Whitehall, which was only large enough for five or six players. The soloists were probably placed nearby, in the organ loft or another gallery, while the rest of the singers were in the choir-stalls on the floor of the chapel. This explains why there is so much antiphonal writing between soloists, choir, and strings in Restoration symphony anthems. They were effectively polychoral works, and modern performances with the musicians in a single mass, are, so to speak, two-dimensional realizations of three-dimensional music.

The first Restoration symphony anthems were probably composed by Henry Cooke—his 'Behold, O Lord, our defender', written for Charles II's coronation on 23 April 1661, is the earliest surviving one that can be dated[38]—but they are fairly rudimentary works, and had little influence on subsequent developments. The composers who wrote most of the symphony anthems in the 1660s and 1670s were Locke, Humfrey, and Blow, and the whole of the front section of Mu. MS 88 was devoted to them—apart from a fragment of the so-called 'Club Anthem', written jointly by Humfrey, Blow, and William Turner.[39] The three anthems by Locke, 'When the son of man shall come in his glory', 'The Lord hear thee in the day of trouble', and 'I will hear what the Lord God will say' (Mu. MS 88, nos. 11–13), are probably the earliest, and represent a transitional type, written, I have argued, in the

[37] For the role of strings in the Restoration Chapel Royal, see Holman, *Four and Twenty Fiddlers*, 397–407.

[38] GB-Bu, MS 5001, 118–25; see H. W. Shaw, 'A Collection of Musical Manuscripts in the Autograph of Henry Purcell and Other English Composers *c.*1665–85', *The Library*, 14 (1959), 126–31.

[39] Humfrey, *Complete Church Music: I*, no. 8.

1660s for cornetts and sackbuts rather than violins.[40] They have verse passages accompanied by continuo in the Restoration manner, but also some extremely old-fashioned solos supported by the instruments in the Jacobean style. We have one completely modern symphony anthem by Locke, 'O be joyful in the Lord, all ye lands',[41] and we have already noted Purcell's debt to 'Be thou exalted Lord', but in general Locke's continuo anthems seem to have been more influential than his symphony anthems.

On the other hand, the five symphony anthems by Humfrey and the four by Blow in Mu. MS 88 represent the mainstream of the genre in the 1660s and 1670s, and would have offered Purcell a range of useful models. Humfrey was an accomplished composer who was capable of considerable intensity in penitential texts and in minor keys. But he worked within a narrow range. Most of the writing in his anthems falls into one of two rhythmic types: the duple-time sections tend to be in the declamatory style, in simple patterns of crotchets and repeated quavers, while the triple-time sections are in a minuet-like rhythm, often with graceful dotted patterns. This dance-like type of writing is often used to excess in Restoration verse anthems, partly, no doubt, because the king could beat time to it, but also because it was the obvious idiom for violins. The instrument was still associated with dance music in England, and composers were reluctant to write for it in any other way, just as the saxophone and the electric guitar have been typecast in this century.

Humfrey was never much interested in counterpoint. The ensemble verse passages and choruses are mostly just in block chords, and there is hardly any melismatic writing. The interest is maintained largely by the harmonies, which are frequently highly expressive, with false relations, chromaticism, and affective chords; yet his harmonic plans are more logical and clearly directed than Locke's, and the chords change at a less furious pace. Humfrey was also good at constructing large structures out of his small building-blocks, frequently repeating part or all of the opening symphony in the middle of the work, and sometimes, as in 'O praise the Lord' and 'O Lord, my God' (Mu. MS 88, nos. 1 and 2), bringing back previously heard vocal sections at the end.[42]

[40] Locke, *Anthems and Motets*, nos. 15, 13, 9; see Holman, *Four and Twenty Fiddlers*, 394–6.

[41] Locke, *Anthems and Motets*, no. 12.

[42] Humfrey, *Complete Church Music: II*, nos. 15, 14.

In the winter of 1664–5 Humfrey was sent by Charles II to study for two years in France and Italy. Much has been made of the possibility that he went to Lully, and was responsible for introducing features of the *grand motet* into the anthem, though we have no idea where he went or what he did, and virtually all the features of his mature style are already present in 'Haste thee O God', a work he wrote before he left England; the words were printed in the 1664 edition of Clifford's *Divine Services and Anthems*.[43] In particular, it starts with a two-section symphony, the first in duple time, the second in triple time, as in the French overture. It is possible that Humfrey encountered the form before going abroad—the violinist John Banister made at least one trip to France in 1660–2—though it must be said that the symphony in this anthem is only French in outline: the pavan-like first section has only a minimum of dotted notes, while the second is a minuet-like dance, not a fugue. One suspects that English composers developed their own form of overture after hearing about the French form rather than seeing or hearing any examples.

John Blow was a much more ambitious and wide-ranging composer than Humfrey, though he did not always maintain his colleague's consistently high standard of workmanship. It is difficult to generalize about the style of his symphony anthems, so varied are they, but one of the greatest, 'Cry aloud, and spare not' (Mu. MS 88, no. 8), reveals several of his preoccupations.[44] One is counterpoint. The symphony is in stark contrast to the type developed by Humfrey. Blow seems to be trying to demonstrate the rival virtues of three national types of contrapuntal writing: traditional English imitation for the first few bars, reminiscent of the full anthem or the viol fantasia, then a burst of brilliant Italianate writing in patterns of semiquavers, and then a triple-time fugal passage with dotted notes in the overture style, the parts entering in order from top to bottom in the French manner. There is also a good deal of counterpoint in the vocal sections: the motifs setting 'Cry aloud, and spare not' and 'lift up thy voice like a trumpet' are worked out in different ways each time they return.

It should not be thought that Blow's symphony anthems are more old-fashioned than Humfrey's just because they are more contrapuntal. Rather than reverse, for the counterpoint in 'Cry aloud, and spare not' is used to create greater contrast—with the

[43] Humfrey, *Complete Church Music: I*, no. 3.

[44] J. Blow, *Anthems III: Anthems with Strings*, ed. B. Wood (MB 64; London, 1993), no. 2.

homophonic passages, and between the contrasted motifs in the contrapuntal passages. It is this greater variety, caused particularly by the introduction of Italianate semiquaver passages, that makes even Blow's early anthems seem so much more modern than Humfrey's, despite the fact that Humfrey's harmony is frequently more logical and tonally directed than Blow's.

Blow is also more concerned than Humfrey to construct taut, unified structures. 'Cry aloud, and spare not' is a beautiful example. It consists of a large verse section, preceded by the symphony and followed by a chorus. The verse is laid out in rondeau form, with the vigorous main passages returning periodically in whole or part, interspersed with episodes of new material. Blow draws attention to the structure by adding a violin obbligato to the episodes, and by taking each one into new harmonic territory. A large unitary structure of this sort, symmetrical in outline but subtly varied in detail, articulated by its themes and harmonies rather than by constant changes of time, was something new in English music, and must have had a profound impact on the young Purcell.

We have twenty-seven complete or substantially complete symphony anthems by Henry Purcell, as well as two fragmentary works, 'If the Lord himself' ZN66 and 'Praise the Lord, ye servants' ZN68. It is much easier to establish a chronology for them than for the continuo anthems. Purcell copied most of the earlier ones, probably more or less in order of composition, into the front portion of GB-Lbl, R.M. 20.h.8, while all the late ones can be dated in one way or another—mostly from annotations in a score in John Gostling's hand at US-Aus.[45]

However, the first two, 'My beloved spake' Z28 and 'Behold now, praise the Lord' Z3, were evidently composed before he started collecting symphony anthems in R.M. 20.h.8. Instead, they survive in early autograph scores in GB-Lbl, Add. MS 30932, the second volume of the 'Flackton Collection', apparently started by Daniel Henstridge, organist of Rochester Cathedral 1673–98 and Canterbury Cathedral 1698–1736; the collection certainly has a Canterbury provenance.[46] The autograph of 'My beloved spake'

[45] F. B. Zimmerman, 'Anthems of Purcell and Contemporaries in a Newly Rediscovered "Gostling Manuscript"', *AcM* 41 (1969), 55–72, and the facs. edn., ed. id., *The Gostling Manuscript* (Austin, Tex., and London, 1977).

[46] Ford, 'Purcell as His Own Editor', 48–9; for Henstridge, see Shaw, *Succession of Organists*, 47–8, 235–6.

must be very early. It gives the work in its unrevised form, and the revised version must be earlier than December 1677, for it appears in the Chapel Royal part-book GB-Lbl, Add. MS 50860—in which Blow is referred to throughout as 'Mr'. 'Behold now, praise the Lord' cannot be dated precisely, but it was probably written in the late 1670s.

'My beloved spake' (ATBB verse, SATB full, four-part strings), a setting of lines from the Song of Solomon, is a bold and confident essay in Humfrey's style. It starts with a minuet-like symphony which Purcell greatly expanded in the revised version, presumably because it seemed too slight to begin a work of more than 300 bars. The vocal sections that follow are largely homophonic in the Humfrey manner, and there are some beautiful passages of expressive harmony, notably at the words 'For lo the winter is past' and 'And the voice of the turtle is heard in our land', when F minor is suddenly substituted for the home key, F major.

The complex 'patchwork' design, with twelve changes of time, also owes much to Humfrey: the symphony is repeated in the middle, and the setting of 'Rise, my love, my fair one, and come away' returns near the end, prompted by a repetition in the text. But Purcell was evidently concerned to widen the scope of the Humfrey type of symphony anthem: there are several new types of writing—including the passage just mentioned, set to delightful bourée-like music. 'My beloved spake' is a notable achievement for a teenager, and it is no accident that the work, with its fresh and sensuous evocation of the spring, has always been one of Purcell's most popular anthems.

'Behold now, praise the Lord' (ATB verse, SATB full, four-part strings), a setting of Ps. 134: 1–3, marks a further departure from Humfrey's style. After the duple-time opening of the symphony, the work is a continuous sweep of dance-like triple time—from the second section of the symphony, through two verse passages and their attached ritornellos, to the fine polychoral doxology. Locke experimented with polychoral effects in 'Be thou exalted Lord', but Blow and Purcell exploited them on a regular basis in symphony anthems, setting the group of soloists in the gallery against the choir below. In 'My beloved spake' there is a rapid exchange between the two groups in the last section, but in 'Behold now, praise the Lord' their phrases overlap, and the strings have independent parts, producing ten-part writing—as the young composer proudly pointed out in the autograph.

The symphony of this anthem is worthy of comment. It is similar in outline to the two-section Humfrey type, but is more complex in its part-writing. In the first section cascading semiquavers are freely mixed with jagged dotted rhythms, the whole propelled forward by dissonant harmony and a series of rapid and unexpected modulations (Ex. 4.7). The harmonic plan, veering through eight or nine keys in twelve bars, recalls Locke, but it is significant that there are no fewer than six 6–4–2 chords in the first section, for sharp dissonance of this sort is a feature of the first sections of French overtures. By the time he wrote it, presumably, Purcell had heard or seen some examples by Lully.

We do not know for sure when Purcell began to copy symphony anthems into R.M. 20.h.8, but it may have been when he began the sequence of secular works at the other end; the first item there is the ode 'Swifter, Isis, swifter flow' Z336, performed in August 1681. Purcell seems to have copied anthems into the volume until the spring of 1685, though another hand added more later. The last autograph anthem is 'My heart is inditing' Z30, written for James II's coronation on 23 April 1685. The datable works in the secular sequence show that he copied in chronological order at a reasonably steady rate. If the same is true of the anthems, then he composed them at an average rate of one every three months over 3½ years. Thus, we can give them tentative dates. 1681: 'It is a good thing to give thanks' Z18 and 'O praise God in his holiness' Z42; 1682: 'Awake, put on thy strength' Z1, 'In thee, O Lord, do I put my trust' Z16, 'The Lord is my light' Z55, and 'I was glad when they said unto me' Z19; 1683: 'My heart is fixed, O God' Z29, 'Praise the Lord, O my soul, and all that is within me' Z47, 'Rejoice in the Lord alway' Z49, and 'Why do the heathen so furiously rage together?' Z65; 1684: 'Unto thee will I cry, O Lord' Z63, 'I will give thanks unto thee, O Lord' Z20, 'I will give thanks unto the Lord' Z21, and 'O Lord, grant the king a long life' Z38; 1685: 'They that go down to the sea in ships' Z57 and 'My heart is inditing'.

Corroboration that these dates are not too wide of the mark is provided by 'They that go down to the sea in ships'. According to Hawkins, it commemorated the occasion when the king, the Duke of York, and the singer John Gostling narrowly escaped shipwreck while in the royal yacht off the Kent coast.[47] Hawkins did not

[47] Hawkins, *General History*, ii. 693.

give the story a date, but added that Charles II did not live to hear the anthem, so it was probably written shortly before the king's death on 6 February 1685. Purcell only copied a few bars in R.M. 20.h.8, perhaps because the impending coronation left him with no time to finish the task. For some reason, too, he

Ex. 4.7. 'Behold now, praise the Lord' Z3, bars 1–12

Ex. 4.7. *cont.*

never copied 'I will give thanks unto thee, O Lord' and 'O Lord, grant the king a long life' (which would have been inappropriate after 6 February), though he entered them in the index; they only survive in secondary sources.

The first two symphony anthems in R.M. 20.h.8 are no great advance on 'My beloved spake' and 'Behold now, praise the Lord', though they are the first of a long series of Purcell works with the distinctive wide-ranging solo bass parts associated with John Gostling. Gostling started his career at Canterbury Cathedral, and became a Gentleman of the Chapel Royal at the end of February 1679.[48] In a letter dated 8 February of that year Thomas Purcell told him: ''tis very likely you may have a summons to appeare among us sooner then you Imagin: for my sonne is Composing wherein you will be chiefly consern'd'.[49] Zimmerman has argued that the work in question was the continuo anthem 'I will love thee, O Lord' ZN67, though its solo bass part does not have the low notes associated with Gostling (sometimes marked with his name in contemporary scores); moreover, the work is found only in peripheral manuscripts, is not particularly distinguished, and could well be spurious.[50]

Thomas Purcell referred to Gostling's low notes in a playful postscript to his letter—'F faut: and Double E lamy are preparing for you'—and the solo bass parts of 'It is a good thing to give thanks' and 'O praise God in his holiness' span more than two

[48] Rimbault, *Old Cheque-Book*, 16–17.

[49] Now in J-Tn, no. 94; reproduced in *Nanki Music Library [Summary Catalogue]* (Tokyo, 1967), 36; see also Cummings, *Henry Purcell*, 28; the letter is a key document in the debate over Purcell's parentage, see Ch. 1, n. 7.

[50] F. B. Zimmerman, 'A Newly Discovered Anthem by Purcell', *MQ* 45 (1959), 302–11.

octaves, from *D* to *e′*, while that in 'Awake, put on thy strength' has the range, *E–e′*. The most spectacular Gostling part is in 'They that go down to the sea in ships', which ranges several times across two octaves illustrating the words 'They are carried up to heav'n, and down again to the deep', though the range of the bass solo in 'I will give thanks unto thee, O Lord' is even wider: at the words 'as for the proud he beholdeth them afar off' Gostling was required to plunge suddenly from *f′* to *C′*.

'Awake, put on thy strength' (AAB verse, four-part strings, with a missing full section) has several novel features. It is the earliest Purcell anthem with a fugal passage instead of a dance as the second section of the symphony, and the earliest with a ground bass. As so often happened, Blow made the innovations—the former, as we have seen, in 'Cry aloud and spare not', the latter in 'O give thanks and call upon his name' (1680 or before)[51]—and Purcell followed. Nevertheless, Purcell linked the fugue and the ground bass in a novel way: the fugue is repeated towards the end of the work, and then its countersubject, a four-bar major-key elaboration of the *passacaglia*, becomes the theme of the ground, the final three-voice Alleluia; the choral conclusion was not copied into R.M. 20.h.8 and is lost. Purcell did not use ground basses as regularly in his anthems as he did in his odes, though the next item in R.M. 20.h.8, 'In thee, O Lord, do I put my trust', opens with an expressive symphony based on the same rising ground as the song 'O solitude, my sweetest choice' Z406, and closes with a vigorous Italianate Alleluia set to a two-bar ground.

As Purcell increased the dimensions of his symphony anthems—one or two towards the end of the sequence in R.M. 20.h.8 run to more than 400 bars—he evidently felt the need to experiment with new methods of unifying the disparate elements, though he never went as far as William Turner, who constructed the whole of his symphony anthem 'Behold now, praise the Lord' on a single ground bass.[52] In 'I was glad when they said unto me' (ATB verse, SATB full, four-part strings) Purcell reversed the normal relationship between the voices and the strings by relating several of the ritornelli to the succeeding verse passages; the normal pattern was for the strings to take up and repeat ideas from the preceding verse passage.

[51] 'O give thanks and call upon his name' is in Blow, *Anthems III*, ed. Wood, no. 4.
[52] Dearnley, *English Church Music*, 224–5.

In 'I will give thanks unto thee, O Lord' (SSATB verse, SSATB full, four-part strings) he went a stage further. The first verse passage consists of two ideas, the first of which comes from the preceding triple-time section of the overture, while the second provides material for the next ritornello. The second verse section also provides material for a succeeding ritornello, but when the chorus enters it returns to the words and the music (somewhat reworked) of the first verse section, and that is followed by a complete repeat of the opening symphony. Repetitions of this sort cease to be effective when the listener begins to be able to predict them, and for this reason Purcell abandoned them in the second half of the anthem. He tended to avoid symmetrical and predictable repetition schemes, though 'Rejoice in the Lord alway' (ATB verse, SATB full, four-part strings), the famous Bell Anthem (so called because the opening symphony is based on a descending octave peal of bells in the bass), is virtually in rondeau form, with many repetitions of the catchy minuet-like theme.

It is not possible to comment on all the symphony anthems in R.M. 20.h.8, fine as most of them are, but mention must be made of the remarkable and little-known 'My heart is fixed, O God' (ATB verse, SATB full, four-part strings). It uses the same device Purcell used so effectively in the 1685 ode 'Why, why are all the Muses mute?' Z343. He leads the listener to think the work is a continuo anthem, for it starts with a verse passage accompanied just by continuo, and the strings only enter when summoned by the voices with the words 'Awake up my glory, awake, lute and harp' (Ex. 4.8). But thereafter, as if to compensate for the absence of an opening symphony, he gives the strings a succession of ravishing minuet-like ritornelli. In many ways, the work is the epitome of the worldly, dance-based symphony anthem of Charles II's reign: it is largely in the 'step tripla', the instruments are just as important as the solo voices, and the choir only makes a brief and perfunctory appearance at the end.

'My heart is inditing' (SSATTBBB verse and full, four-part strings), the great anthem for James II's coronation, ends the autograph sequence of anthems in R.M. 20.h.8, and also marks the beginning of the end of the symphony anthem. James II did little to disguise his Catholic sympathies, and attended the queen's Catholic chapel at St James's, and then his new Catholic chapel in Whitehall, opened on Christmas Day 1686. The regular payments to string-players for service in the Anglican chapel were discontin-

Ex. 4.8. 'My heart is fixed, O God' Z29, bars 37–42

ued, though violinists continued to attend when Princess Anne was present.[53] A number of symphony anthems in the Gostling score-book have dates in James II's reign, including a group of five by Purcell: 'Behold, I bring you glad tidings' Z2 ('For Christmas day | 1687'), 'Blessed are they that fear the Lord' Z5 ('Jan: 12. 1687. | For the Thanksgiving—| Appointed in London | & 12 miles round, upon her | Majesties being w(i)th Child. | & on the 29 following, over England.'), 'Praise the Lord, O my soul, O Lord my God' Z48 ('1687'), 'Thy way, O God, is holy' Z60

[53] Holman, *Four and Twenty Fiddlers*, 411.

('1687'), and 'O sing unto the Lord' Z44 ('1688').[54] 'Thy way, O God, is holy' is printed in *Works II*, 32 as a continuo anthem, but it appears in the Gostling score with simple passages for two violins and bass. Conversely, 'O give thanks unto the Lord' Z33 ('1693') is in *Works II*, 29 as a symphony anthem, though there is no opening symphony, and the short three-part ritornelli are reduced for organ in most sources; Gostling copied it into the continuo anthem section of his book.[55]

'Behold, I bring you glad tidings' (ATB verse, SATB full, four-part strings), a setting of familiar words from St Luke's gospel, reflects Purcell's increasing interest in the Italian style, and in particular, in Draghi's setting of Dryden's St Cecilia ode 'From harmony, from heavenly harmony'. This influential work (which will be discussed more fully in Ch. 5), must have been fresh in Purcell's ears, for it had been first performed just a month earlier, on 22 November 1687. Instead of opening his symphony with the usual passage of dotted rhythms, Purcell followed Draghi in writing a short passage of solemn repeated chords in a rising pattern; the second section is not a minuet-like dance, but a vigorous duple-time fugue, similar in style to the canzonas of the trio sonatas. The first verse passage, too, is an Italianate recitative accompanied by the strings, as in several of the solos of the 1687 ode, and in the following ATB verse the strings are not confined to a concluding ritornello, but continually answer the voices, as in Italian operatic arias.

More generally, we can hear the influence of the 'fam'd Italian masters' in the harmonies. There is less emphasis on intense passing dissonance and on rapid, surprising modulations, and more on logical long-range harmonic planning. The fugue of the symphony, for instance, uses the standard plan popularized by Vivaldi and his contemporaries, modulating to the dominant and then back to the tonic by way of two minor keys. However, the most remarkable feature of this fine anthem is its sense of drama. It is hard not to think of the solo bass as the Angel, delivering the good news, or the continual exchanges between soloists and choir at the words 'Glory to God on high, and on Earth peace' as the dialogue between the heavenly host and the shepherds—the more so since the soloists would have been placed aloft in a gallery, the choir on the floor of the chapel (Ex. 4.9).

Italianate features are even more marked in Purcell's last sym-

[54] Zimmerman, 'Anthems of Purcell and Contemporaries', 63–4.

[55] Ibid. 68.

Ex. 4.9. 'Behold, I bring you glad tidings' Z2, bars 189–202

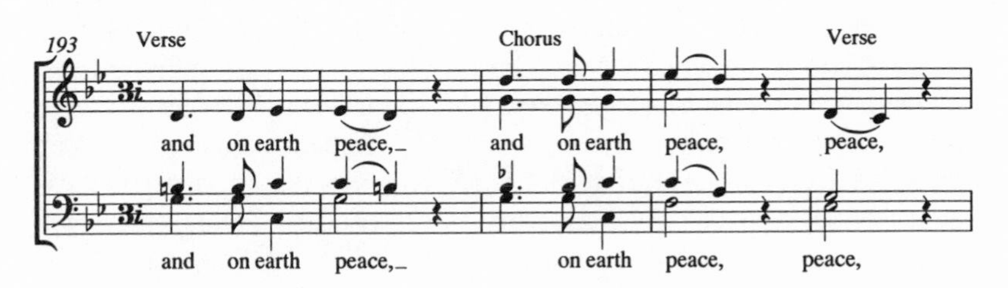

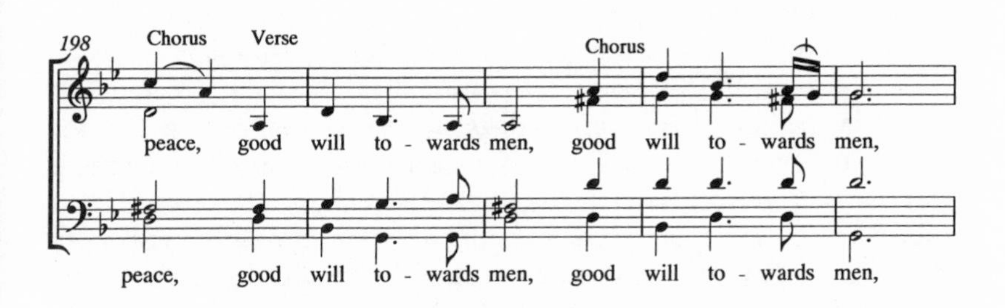

phony anthems. The popular 'O sing unto the Lord' starts, like 'Behold, I bring you glad tidings', with an 'Italian overture', and like it has a passage of recitative for solo bass accompanied by the strings. But Purcell is more concerned in this anthem to divide the work into separate movements, distinguishing them by key as well as by scoring and the type of motion used. This was a trend all over Europe in the 1680s and 1690s, and was an inevitable consequence of the drive, fuelled by the development of standard modulation schemes, to create longer and longer works. 'O sing unto the Lord' effectively consists of seven separate movements, the third and fifth of which are in related minor keys (D minor and F minor) to the home key (F major). The fact that they are placed either side of a central ground bass movement, the duet 'The Lord is great', increases the sense of arch-like symmetry.

Symmetry is also important in 'My song shall be alway' (B or S verse, SATB full, four-part strings), a virtuoso solo anthem with only a minimal choral contribution. The work is usually dated 1687 or 1688, and has sometimes been associated with the Italian castrato Giovanni Francesco Grossi, who was in England for a few months in 1687. But a score in the hand of the Oxford musician Francis Withey dates the work 9 September 1690, and on that day William III entered Windsor in triumph after the siege of Limerick, an occasion that could well have called for a celebratory anthem.[56] A part-autograph set of instrumental parts, which might have been used in the first performance, survives in GB-Och, Mus. MSS 1188–9; Withey was a singing man at Christ Church.[57] Like many Purcell anthems, 'My song shall be alway' is divided into two halves by a repeat of the opening symphony (another 'Italian overture') halfway through. But the sense that the work is a diptych in two panels is increased by the fact that the same short choral Alleluia ends each half, and that each half contains the equivalent of a recitative followed by an aria, a short intermediate ritornello, as well as an excursion into the tonic minor. As I observed at the end of Ch. 2, symmetry of this sort is characteristic of English Baroque art.

By the time Purcell wrote 'My song shall be alway' the practice of using groups of instrumentalists in the Chapel had ended for good. On 23 February 1689 Queen Mary issued an order banning instruments in the chapel, and in 1691 William III ordered that the 'King's Chappell shall be all the year through kept both morning and evening with solemn musick like a collegiate church', which carries the implication that violins were to be excluded, for they were not used in collegiate foundations outside the court.[58]

Thereafter, church music with orchestra was only written for special occasions. Purcell's Te Deum and Jubilate in D major Z232

[56] Holman, *Four and Twenty Fiddlers*, 406.

[57] B. Wood, 'A Newly Identified Purcell Autograph', *ML* 59 (1978), 329–32.

[58] GB-Lpro, RG8/110, fos. 24^r–25^v, a reference kindly brought to my attention by Paul Hopkins; H. C. de Lafontaine (ed.), *The King's Musick: A Transcript of Records Relating to Music and Musicians (1460–1700)* (London, 1909; repr. 1973), 407.

(SSAATB verse, SSATB full, two trumpets, four-part strings) is a good example. Thomas Tudway wrote that the work was 'th(a)t Noble Composition, the first of its kind in England, of Te Deum, & Jubilate, accompanied w(i)th instrumentall music; w(hi)ch he compos'd principally against the opening of St Pauls, but did not live till that time'.[59] Z232 does seem to have been the first English Te Deum and Jubilate with orchestra, the inspiration for similar works by Blow, Croft, and Handel. But Tudway seems to have been mistaken about its date, for the title-page of the printed score of 1697 states that it was 'Made for St. Caecilia's Day, 1694.'; it was probably first heard at the service in St Bride's church that preceded the main St Cecilia celebrations at Stationers' Hall.

The Te Deum and Jubilate were probably Purcell's most admired compositions in the eighteenth century. Tudway went into raptures:

> there is in this Te Deum, such a glorious representation, of the Heavenly Choirs, of Cherubins, & Seraphins, falling down before the Throne & singing Holy, Holy, Holy &c As hath not been Equall'd, by any Foreigner, or Other; He makes the representation thus; He brings in the treble voices, or Choristers, singing, To thee Cherubins, & Seraphins, continually do cry; and then the Great Organ, Trumpets, the Choirs, & at least thirty or forty instruments besides, all Joine, in most excellent Harmony, & Accord; The Choirs singing only, the word Holy; Then all Pause, and the Choristers repeat again, continually do cry; Then, the whole Copia Sonorum, of voices, & instruments, Joine again, & sing Holy; this is done 3 times upon the word Holy only, changeing ev'ry time the Key, & accords; then they proceed altogether in Chorus, w(i)th, Heav'n, & Earth are full of the Majesty of thy glory; This most beautifull, & sublime representation, I dare challenge, all the Orators, Poets, Painters &c of any Age whatsoever, to form so lively an Idea, of Choirs of Angels singing, & paying their Adorations (Ex. 4.10).

We can see what he meant, and doubtless the sound of 'the Great Organ, Trumpets, the Choirs, & at least thirty or forty instruments besides' was novel and thrilling to Tudway and his contemporaries in itself. But it is hard not to notice that most of the sections of the Te Deum are extremely short-winded, and that they do not really add up to a coherent whole. The Jubilate, with less text to get through, is rather better in this respect, though it is easy to understand why Croft greatly expanded the dimensions of his Te Deum and Jubilate, creating memorable extended separate movements for the more expressive sections of text, such

[59] Hogwood, 'Thomas Tudway's History of Music', 45.

Ex. 4.10. Te Deum in D major Z232, bars 51–9

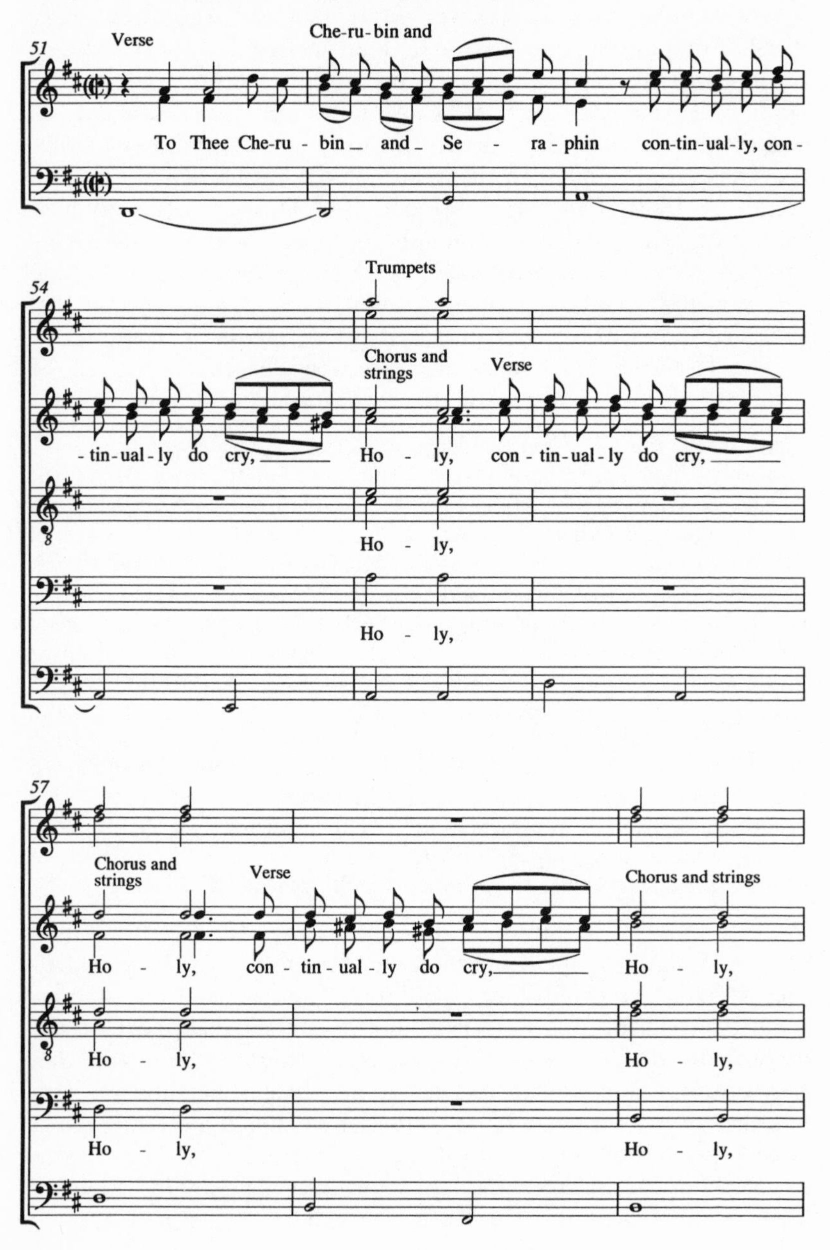

as 'The father of an infinite majesty' and 'Vouchsafe, O Lord, to keep us this day'.[60] For the same reason William Boyce rewrote Purcell's Te Deum in 1755, adding longer instrumental passages, repeating sections, and doubling the work in length.[61]

Purcell's last church music was written for Queen Mary's funeral, in Westminster Abbey on 5 March 1695. It consists of the funeral sentence 'Thou know'st, Lord, the secrets of our hearts' Z58C, and the March and Canzona Z860 for four 'flat' (i.e. slide) trumpets. According to the main source, GB-Ooc, MS U.a.37, the March was 'sounded before her Chariot', apparently in the procession (though experiments with modern copies of flat trumpets suggest that the players would have had to have been stationary[62]), while the Canzona was 'sounded in the Abby after the Anthem'. The March was later used in Shadwell's play *The Libertine* Z600/2a (see Ch. 6).

Bruce Wood has argued that 'Thou know'st Lord' was written by Purcell to fit into Thomas Morley's Burial Service, which had been used for state funerals before the Civil War.[63] Several manuscripts preserve the Purcell with the Morley, which seems to have been transmitted in the Restoration period with the 'Thou know'st Lord' sentence missing. William Croft later incorporated Purcell's setting into his own Burial Service, for reasons, he wrote in *Musica sacra*, i (London, 1724), 'obvious to every Artist'.[64] Tudway wrote that it was 'accompanied w(i)th flat Mournfull Trumpets', and it is appropriate to end this survey of Purcell's church music with his description of the effect that his matchless music made at its first performance:

> I appeal to all th(a)t were p(re)sent, as well such as understood Music, as those th(a)t did not, whither, they ever heard any thing, so rapturously fine, & solemn, & so Heavenly, in the Operation, w(hi)ch drew tears from all; & yet a plain, Naturall Composition; w(hi)ch shews the pow'r of Music, when tis rightly fitted, & Adapted to devotional purposes . . .[65]

Within the year it was performed again, at Purcell's own funeral, in Westminster Abbey on 26 November 1695.

[60] W. Croft, *Te Deum*, ed. H. W. Shaw (London, 1979); I am grateful to Lucy Roe for providing me with a transcription of Croft's Jubilate.

[61] W. H. Cummings, 'The Mutilation of a Masterpiece', *PMA* 30 (1903–4), 113–14.

[62] I am grateful to Crispian Steele-Perkins for this information.

[63] Wood, 'First Performance of Purcell's Funeral Music'.

[64] W. Croft, *The Burial Service*, ed. B. Wood (Croydon, 1985).

[65] Hogwood, 'Thomas Tudway's History of Music', 28–9.

V
ODES

BY and large, Purcell's odes have received short shrift in modern times. The texts are routinely dismissed as inept and embarassingly sycophantic, and attention has largely been confined to the three works currently available in miniature score: 'Welcome to all the pleasures' Z339, 'Hail, bright Cecilia' Z328, and 'Come, ye sons of art, away' Z323.[1] The original volumes in *Works I* containing the odes were published between 1878 and 1926, and have been out of print for decades. They are only now being replaced by volumes in *Works II*, and some first-rate pieces are still virtually unknown. Purcell's odes are important because nearly all of them can be dated precisely—unlike most of the music discussed so far in this book. The sequence of twenty-four surviving works offers an unparallelled opportunity to observe successive changes in Purcell's style between 1680 and the last few months of his life.

The choral and orchestral ode seems to have been an English invention.[2] It used to be thought that Restoration composers borrowed the form from France, just as it was thought that the symphony anthem was derived from the *grand motet*. But French composers never developed a choral and orchestral genre equivalent to the English ode, and Lully only established the practice of glorifying the king in operatic prologues in the 1670s. The English court ode had antecedents before the Civil War, such as Orlando Gibbons's 'Do not repine, fair sun', written to welcome James I to Edinburgh in 1617, and Ben Jonson's text for 'A New-yeares-Gift sung to King Charles, 1635[/6]', for which the music is lost.

We do not know when the first Restoration odes were written. The genre may have come into being as a means of celebrating 29 May, the public day of thanksgiving ordered by Parliament in

[1] H. Purcell, *Welcome to all the Pleasures, Ode for St. Cecilia's Day 1683*, ed. W. Bergmann (London, 1964); id., *Ode for St. Cecilia's Day (1692)*, ed. M. Tippett and Bergmann (London, 1955); id., *Come ye Sons of Art*, ed. Tippett and Bergmann (London, 1951).

[2] For a history of the ode, see R. McGuinness, *English Court Odes 1660–1820* (Oxford, 1971), 1–61.

1660 to celebrate the double anniversary of Charles II's birthday and his return to London, though the earliest surviving odes are Henry Cooke's 'Good morrow to the year' and Matthew Locke's 'All things their certain periods have', written for the New Year 1666. We also have the texts but not the music of 'A Pastorall Song, to the King on Newyearesday. An(n)o Domini 1663[/4]', an adaptation of part of Jonson's 1636 ode, set by Nicholas Lanier, and Lanier's 'Come loyal hearts, make no delay', also set by Locke for the New Year 1666. The earliest surviving odes for 29 May are 'Rise thou, best and brightest morning' and 'Come, we shepherds', undated but set by Cooke (d. 13 July 1672), though a short part-song by Locke, 'Welcome, welcome, royal May', published in Playford's *Catch That Catch Can; or, The Musical Companion* (1667), pp. 118–20, is drawn from a poem by Alexander Brome 'On the King's returne', published in 1661.[3] A third type, the welcome song or ode, written to greet the monarch on his return to Whitehall after a period of absence, came into being in 1680 with Purcell's first ode 'Welcome, vicegerent of the mighty king' Z340.

We know remarkably little about the circumstances in which court odes were performed. The early odes are either just for voices and continuo, or for voices, two violins, and continuo, and may therefore have been given in the Privy Chamber by a small group of musicians from the Private Music, though the only description from this period, in the account of the Grand Duke of Tuscany's visit to England in 1669, mentions that the king dined in public in the Banqueting House at Whitehall on his birthday, and that dinner was 'enlivened with various pieces of music, performed by musicians of the king's household'.[4]

By the early 1670s, when Pelham Humfrey was writing court odes, the resources called for had expanded considerably. His three surviving odes, 'See, mighty Sir, the day appears' (1 January 1672), 'When from his throne the Persian god appears' (29 May 1672), and 'Smile, smile again twice happy morn' (?29 May 1673), are considerably longer and more complex works than Cooke's, and call for three, four, or five solo voices and chorus with four-part strings and continuo. They were probably performed in a public part of the palace, such as the Hall Theatre at Whitehall;

[3] Harding, *Matthew Locke*, 38.

[4] L. Magalotti, *Travels of Cosmo the Third, Grand Duke of Tuscany, through England, during the Reign of King Charles the Second (1669)* (London, 1821), 364–5.

indeed, payments survive for making alterations to the Hall Theatre in May 1678 and the beginning of 1685, perhaps for the performances of Blow's 1678 birthday ode 'The birth of Jove', and his 1685 New Year ode 'How does the new-born infant year rejoice?'.[5] The earliest description of the performance of an ode, on 29 May 1681, mentions that 'At His Majesty's up-rising a Song was sung in the Privy Chamber, concerning the Birth, Restauration, and Coronation, much to the satisfaction of His Majesty', but the work, Blow's 'Up, shepherds, up', is apparently scored just for voices and continuo, and is thus untypical of odes of the period.[6]

There does not seem to have been a fixed time and place for the performance of court odes. We have already seen that the Privy Chamber in the morning was one venue, and the Banqueting House during dinner may have been another. On New Year's Day 1690 the court chronicler Narcissus Luttrell recorded that 'the king and queen came to Whitehall, where many of the nobility and gentry came to wish them a happy new year; and there was a great consort of musick, vocal and instrumental, and a song composed by the poet laureat'.[7] On this occasion the ode followed formal New Year greetings, so it was probably performed in the morning. The same thing probably happened on 1 January 1696, when courtiers went to call on the king and queen at Kensington Palace. According to a newspaper report, 'all the Nobility, Judges, Gentlemen, &c., went to Kensington and wished His Majesty many joyful and happy years, the Trumpets, Kettle-Drums, &c., made that performance, as did also his Majesty's Musicianers, and a curious Ode, being composed on this occasion was sung and plaid to'.[8]

On the other hand, at least one of the odes for James II was performed in the evening. Samuel Pepys wrote to Sir Robert Southwell on 10 October 1685 that 'To night wee have had a mighty Musique-Entertainment at Court for the welcomeing home the King and Queene.'[9] A few days later, on 15 October, John Evelyn was at court for the king's birthday celebration, evidently delayed or prolonged from the 14th, where he heard music

[5] Holman, *Four and Twenty Fiddlers*, 422.

[6] McGuinness, *English Court Odes*, 10.

[7] N. Luttrell, *A Brief Historical Relation of State Affairs from September 1678 to April 1714* (Oxford, 1857; repr. 1969), ii. 1.

[8] McGuinness, *English Court Odes*, 52.

[9] S. Pepys, *Letters and the Second Diary of Samuel Pepys*, ed. R. G. Howarth (London, 1932), 171.

performed before a ball, an evening event: 'Being the Kings birthday, was a solemn Ball at Court; And Musique of Instruments & Voices before the Ball: At the Musique I happen(ed) (by accident) to stand the very next to the Queene, & the King, who ta(l)ked with me about the Musick.'[10] Another indication of the performance of an ode in the evening is the order to the Lord Steward's department to deliver '12 White wax lights' for 'the singing before his Ma(jes)tie' on New Year's Day 1686/7.[11]

As a literary genre, the Restoration ode was heavily dependent on Classical models. Horace's odes had been translated, imitated, and set to music during the Renaissance, and there was a fashion for Pindar in seventeenth-century England. Abraham Cowley published a collection of *Pindarique Odes, Written in Imitation of the Stile and Manner of the Odes of Pindar* in 1656, which includes translations of two by the Greek poet, as well as Horace's tribute to him, and a number of original works in the same style by Cowley himself, including an ode 'To the NEW YEAR' (pp. 34–5), 'Great Janus, who dost sure my mistress view'. It does not seem to have been set by a Restoration composer—its tone is too intimate and reflective for a court celebration—but it shares many features with the odes that Purcell set. The subject-matter is serious and the thought sustained, yet the tone is conversational and informal, and the syntax simple and direct. It may not be great poetry, but it is ideal for setting to music. It falls into stanzas of irregular length, providing the composer with ready-made divisions into a varied series of solos and chorusses. The metre is also irregular, as is the rhyme scheme, which allows the composer to vary the length of his phrases with ease, and to use different time signatures and rhythmic patterns for the various movements.

The authors of Restoration court odes rarely matched the quality of Cowley's thought, and it is as easy to smile at their conventional images and far-flung similes as it is to cringe at their fawning adulation of all-too-fallible monarchs. But we must remember that the Pindaric ode was a formal, public type of poetry (most of Pindar's odes honour the winners of major sporting events such as the Olympic Games), and there is no sign that anyone at the time thought the texts of court odes excessively sycophantic. They only began to seem so as the institution of the monarchy declined in the course of the eighteenth century.

[10] Evelyn, *Diary*, iv. 480. [11] Ashbee, *RECM* v. 233.

Jonathan Swift was articulating a more modern and cynical sensibility when he wrote in his 'Directions for making a Birth-day Song' of 1729:

> Your encomium, to be strong,
> Must be apply'd directly wrong.
> A tyrant for his mercy praise,
> And crown a royal dunce with bays:
> A squinting monkey load with charms,
> And paint a coward fierce in arms.
> Is he to avarice inclin'd?
> Extol him for his gen'rous mind . . .[12]

In later times the texts of court odes were routinely provided by the Poet Laureate, but that does not seem to have been the case in the Restoration period. Apart from the ones by Jonson and Lanier mentioned earlier, and 'See, mighty Sir' and 'When from his throne', which Robert Veel wrote for Humfrey,[13] they are all anonymous until 1684, when Thomas Flatman wrote the New Year ode 'My trembling song, awake, arise' for Blow, and the welcome song 'From those serene and rapturous joys' for Purcell. To judge from their quality, we can safely say that John Dryden was not responsible for any of the anonymous odes produced during his term of office as Poet Laureate (1668–89), though he wrote the St Cecilia odes in 1687 and 1697.

The arch-Whig Thomas Shadwell replaced the Catholic convert Dryden at the accession of William and Mary, and wrote a number of court odes, including the 1689 ode for Mary's birthday (30 April), 'Now does the glorious day appear', set by Purcell. But Shadwell did not have a monopoly. Thomas D'Urfey wrote several odes, including 'Arise, my Muse', set by Purcell for 30 April 1690, while Charles Sedley provided Purcell with one of the finest texts, 'Love's goddess sure was blind', the ode for 30 April 1692. Nahum Tate, who became Laureate after Shadwell's death in 1692, seems to have written a higher proportion of the odes, though there are still a few by others. Matthew Prior wrote his 'Hymn to the Sun', 'Light of the world and ruler of the year', for the New Year 1694, but Purcell's music does not survive, and

[12] McGuinness, *English Court Odes*, 62.
[13] P. Dennison, *Pelham Humfrey* (Oxford, 1986), 76.

may never have existed; it was apparently replaced at short notice by Motteux's 'Sound, sound the trumpet', set by Blow.[14]

A similar pattern can be seen in the choice of composers for court odes. In the eighteenth century the task was routinely undertaken by the Master of the Music, successively John Eccles, Maurice Greene, William Boyce, John Stanley, and William Parsons, but it was shared by a number of composers in the Restoration period, and as late as 1713 and 1714 birthday odes were written by Handel and William Croft. In fact, the Restoration Masters of the Music were not the obvious people to write the odes. The Catalan violinist Louis Grabu (Master 1666–74) seems to have been a fine orchestral trainer and wrote some good orchestral music in the French style, but he was not at his best setting English or writing large-scale concerted works, while Nicholas Staggins (Master 1674–1700) was evidently not much of a composer; he wrote some court odes, but they are lost, as are virtually all his sizeable works.[15]

In fact, the Restoration odes were mostly written by the Master of the Children of the Chapel Royal rather than the Master of the Music: we have four by Henry Cooke (Master of the Children 1660–72), three by Humfrey (Master 1672–4), and more than twenty by Blow (Master 1674–1708). Blow wrote all the surviving New Year and birthday odes from 1674 to the end of James II's reign, apart from two undated birthday odes by William Turner. Purcell was brought into the process when the series of welcome songs started in 1680. All his court odes before 1689 are of this type except one, 'From hardy climes and dangerous toils of war' Z325, written for a royal wedding, though it is possible that the three he wrote to welcome James II to Whitehall after his summer progress also served as odes for the king's birthday, 14 October (see below). The practice of performing welcome songs seems to have been discontinued after the Glorious Revolution, perhaps because there were now two royal birthdays to celebrate. Instead, Purcell was given the job of writing odes for Mary's birthday,

[14] McGuinness, *English Court Odes*, 21, 52; the text is headed 'HYMN to the SUN. Set by Dr. PURCEL, And Sung before their MAJESTIES On New-Years-Day, 1694' in the 1718 edition of Prior's poems, but 'An Ode to the Returning Sun, Intended to be Sung before Their late Majesties, on New-Year's-Day $169\frac{3}{4}$, (but here printed with Alterations; as it was performed lately at a Consort of Musick, by the most Eminent Masters.)' in the 1707 edition; see M. Prior, *Poems on Several Occasions*, ed. A. R. Waller (Cambridge, 1905), 22–4, 344.

[15] For Grabu and Staggins, see Holman, *Four and Twenty Fiddlers*, 293–304.

while Blow continued to write the New Year odes. All the odes for William III's birthday, including Staggins's for 1693, 1694, and 1697, are lost, with the exception of Blow's for 1692 and 1697 (a second ode), which survive incomplete.

We are lucky to have as many court odes as we do, for the sources are meagre. None were printed, and some survive in only one copy. All the performing material was probably lost when Whitehall was burnt down in 1698; Watkins Shaw has argued that GB-Bu, MS 5001, a collection of autograph scores of anthems and odes by Cook, Humfrey, Blow, Turner, and Purcell, was rescued on that occasion.[16] The largest collection of odes is the second portion of GB-Lbl, Add. MS 33287, a score copied apparently in the 1680s by an unknown scribe in the court milieu; it contains twenty-six works ranging in date from Humfrey's two odes of 1672 to Blow's 'Ye sons of Phoebus' (1 January 1688). Add. MS 33287 contains nine odes by Purcell, but the main source of his early works in the medium is the reversed portion of the autograph score-book GB-Lbl, R.M. 20.h.8, which contains all but one of his court odes from the reigns of Charles II and James II in a chronological sequence with other vocal music, as well as a part-autograph copy of 'Celestial music did the gods inspire' Z322, and non-autograph copies of three of the other odes from 1689–90.

It has often been said that the idiom and form of the early court ode were derived from the Restoration symphony anthem, and that the two genres gradually developed in different directions. In fact, there were marked differences between them in the 1660s, which diminished to some extent in the 1680s as Purcell and Blow began to introduce anthem-like contrapuntal writing into the ode, and then widened again in the 1690s as the ode expanded in size, and began to be scored with wind and brass instruments as well as strings. At first, the main difference was the extent to which dance idioms were used. Humfrey's anthems mostly consist of duple-time declamatory passages and dance-like movements in minuet rhythm. But the declamatory passages are largely absent in his odes, and there is a much wider selection of dance-like material. In 'See, mighty Sir, (GB-Lbl, Add. MS 33287, fos. 69^{v}–71^{v}), for instance, there are only two brief declam-

[16] H. W. Shaw, 'Collection of Musical Manuscripts'.

atory passages; the rest of the vocal sections are cast in a variety of dance rhythms: bourée, minuet, gavotte, and jig.

Restoration composers were doubtless aware that there were Classical precedents for setting their odes largely as dance suites. In ancient Greece odes performed as part of public ceremonial were accompanied by dancing. In Greek tragedy the chorus performed in the *orchestra* or 'dancing area' of the theatre, and sang and danced odes at the end of plays. We have no direct evidence that the sections of Restoration odes in dance forms would have been danced, though, as we have seen, some of them were performed in the Hall Theatre at Whitehall, where there would certainly have been sufficient space for dancing. Also, it may be significant that several of Purcell's odes contain extended dance movements whose function is difficult to explain except as vehicles for dancing. 'Sound the trumpet, beat the drum' Z335 contains the Chaconne in F major later used in *King Arthur* Z628/1a, while 'Who can from joy refrain?' Z342 ends with a choral and orchestral chaconne similar to 'Triumph victorious Love' Z627/38, the 'grand dance' in the Act V masque of *Dioclesian*, or the passacaglia danced and sung in Act IV of *King Arthur* Z628/30.

Blow probably took over the composition of court odes after Humfrey's death in 1674, though we do not have any by him that can be dated before 1678, and even then there is scope for argument over the exact dates. However, there seem to be four odes by Blow that are earlier than the first by Purcell: 'Dread Sir, the prince of light' (Add. MS 33287, fos. 75^{r}–78^{r}, ?1 January 1678), 'The birth of Jove' (ibid., fos. 58^{r}–62^{v}, ?29 May 1678), 'Great Janus, tho' the festival be thine' (ibid., fos. 53^{r}–57^{v}, ?1 January 1679), and 'The new year is begun' (ibid., fos. 93^{r}–96^{r}, ?1 January 1680).[17]

In these works Blow mainly conformed to the pattern established by Humfrey. They are typically scored for ATB or SATB solo voices, SATB choir, four-part strings, and continuo, and consist of a patchwork of short sections, mostly in dance rhythms, with the solos often leading to ritornelli, or to repetitions of their words and music by the chorus. There are, however, some novel features. From the first, Blow preferred to cast the second sections of his overtures as fugues, though in 'Dread Sir, the prince

[17] The evidence for dating these works is presented in McGuiness, *English Court Odes*, 15–16, 45–6.

of light' he retained the division into two repeated sections found in the homophonic second sections of Humfrey's overtures. Blow followed Humfrey in setting a good deal of his early odes to minuet-like triple time, but he also began to develop the type of suave and melodious duple-time writing that was to become so fashionable in the 1680s. The choruses are still largely homophonic and syllabic, in four parts with the strings assumed to be doubling the voices, though in 'Great Janus, tho' the festival be thine' they are in six parts, for SSATB with an independent first violin.

Purcell's first ode, 'Welcome, vicegerent of the mighty king', is also a work of this type. It is headed 'A Song to Welcome home his Majesty from Windsor 1680' in John Walter's score, GB-Lbl, Add. MS 22100, the only contemporary copy; according to Luttrell, Charles II arrived back at Whitehall that year on 9 September.[18] It is scored for SSATB solo, SATB choir, four-part strings, and continuo, and is a patchwork of more than a dozen short sections, mostly based on minuet, gavotte, or bourée rhythms. There are some charming moments—as when the chorus echoes each phrase of the tenor soloist in the verse 'Your influous approach our pensive hope recalls', or when two choirboys warble the gavotte 'When the summer, in his glory' in sixths and thirds—and there is one notable novelty: Purcell combines the overture with the first chorus by superimposing voice parts on to the instrumental lines for the repeat of the minuet-like second section, producing independent seven- or eight-part writing for voices and instruments, perhaps for the first time in a court ode (Ex. 5.1). But the problem with basing a concerted work of this sort largely on a patchwork of dance patterns is that there is an uncomfortable disparity between the sizeable scale of the overall work, dictated by the lengthy text, and the tiny size of its component parts. The ear soon tires of a confusing and seemingly endless succession of short repeated phrases, articulated only by sudden changes of rhythmic and harmonic direction.

Purcell and Blow were evidently aware of this problem (which is common to much northern European music of the period), and began to solve it in their next few court odes by reducing the number of sections, and by writing longer, more self-contained, and more developed solo movements. These movements could no longer be organized effectively just by using rhythmic and

[18] Luttrell, *Brief Relation*, i. 54.

Ex. 5.1. 'Welcome, vicegerent of the mighty king' Z340, bars 49–62

Ex. 5.1. *cont.*

56
[tr]
That made and go - verns, and go - verns ev' - ry
That made and go - verns, and go - verns ev' - ry
That made and go - verns, and go - verns ev' - ry
60
thing,
thing,
thing,

harmonic patterns derived from dance music, and so Blow and Purcell turned to ground basses. We saw in Chapter 2 that English composers began to write songs on grounds around 1680, initially mostly using patterns derived from the triple-time *ciaccona* and *passacaglia*. But grounds of this sort are rarely found in odes; instead, Blow and Purcell preferred to devise their own duple-time basses, typically in busy patterns of quavers outlining an ascending or descending sequence of harmonies.[19] The first surviving example seems to be 'Of you, great sir, our Druids spake' from Blow's 'Great Sir, the joy of all our hearts' (?1 January 1681, Add. MS 33287, fos. 112^{r}–117^{v}), though Purcell had already used a similar pattern in a short ritornello for continuo instruments at the end of 'Your influous approach' in 'Welcome, vicegerent'; it is presumably intended to evoke the sound of Apollo's lyre, mentioned in the previous verse.

Purcell's first ode to include a ground bass is his second welcome song, 'Swifter, Isis, swifter flow' Z336. This work is a notable advance on 'Welcome, vicegerent', and was chosen by the composer to open R.M. 20.h.8; it may be the first ode he felt worth preserving. He headed it 'A Welcome Song in the Year 1681 For the King', and it is often assigned to 12 October of that year, when Charles II returned to Whitehall from Newmarket. But the charming text describes the king travelling by boat down the Thames (called the Isis in its upper reaches) to Westminster, and the occasion is therefore more likely to have been his return from Windsor on 27 August—which probably would have been undertaken on the Thames.[20] The text, pastoral in tone and soaked in watery images, perhaps encouraged Purcell to introduce wind instruments for the first time in a court ode; the bass solo 'Land him safely on her shore' is accompanied by two recorders and continuo, while the ground bass, the tenor solo 'Hark! just now my list'ning ears', ends with a ritornello for oboe and strings imitating bells.

Baroque oboes and recorders were probably introduced into England by a group of French wind-players who arrived in London in 1673; three of them seem to have been at court until about 1682.[21] The two remarkable anthems Blow wrote with wind

[19] See R. McGuinness, 'The Ground-Bass in the English Court Ode', *ML* 51 (1970), 118–40, 265–78.
[20] Luttrell, *Brief Relation*, i. 118.
[21] See Holman, *Four and Twenty Fiddlers*, 343–4; 407–11.

instruments around 1680, 'Lord, who shall dwell in thy tabernacle?' and 'Sing unto the Lord, O ye saints of his', show that they were well equipped and versatile, well able to change rapidly between a range of recorders and double reed instruments.[22] Thus their contribution to 'Swifter, Isis' may not have been limited to the parts Purcell wrote specifically for them; it is likely that they doubled the string parts with oboes and bassoon in some or all of the tutti sections.

In 'Swifter, Isis' the process of lengthening the movements and reducing their number is already advanced. The work is about the same length as 'Welcome, vicegerent', but is divided into only eight major sections. Moreover, there is now a greater sense of order and coherence. This is partly because the sections are placed in a logical and more wide-ranging circle of keys: G major, G minor, C major, C major, A minor, F major, D minor, and G major. 'Welcome, vicegerent' still uses a conservative harmonic plan, oscillating between C major and C minor with a single excursion to G minor and G major. Also, the sections of 'Swifter, Isis' are placed in an arch-like pattern, with two vocal minuets placed either side of a declamatory bass solo, and two matching movements, the ground bass 'Hark! just now my list'ning ears' and the duet 'The king, whose presence like the spring' (a 'pseudo-ground'; see below), placed either side of them. The work opens and closes with sequences of sections based largely on dance patterns, but now constructed in a more complex way. In the opening sequence the minuet-like fugue of the overture leads directly to the opening solo, which also uses its descending theme, and then to the first chorus, which is based on it, as in 'Welcome, vicegerent'; the minuet rhythm is continued in a ritornello, a second chorus (apart from a short duple-time introduction), and a final ritornello. As Purcell got older he tended to replace complexes of this sort with separate movements, but he never tired of arranging movements in intricate and quasi-symmetrical sequences.

Purcell headed his next ode, 'What, what shall be done in behalf of the man?' 'A Welcome Song for his Royall Highness at his return from Scotland in the yeare 1682'. It was written to welcome the Duke of York (the future James II) back to London after his time as High Commissioner for Scotland. James had been given the post in 1679 as an alternative to exile during the

[22] J. Blow, *Anthems II: Anthems with Orchestra*, ed. B. Wood (MB 50; London, 1984), nos. 3, 6.

Popish Plot crisis, and was allowed to return in the spring of 1682. In March he sailed from Leith to Yarmouth to meet the king at Newmarket. They returned to London on 8 April, and in early May James started back to Scotland to fetch his family, but on the morning of the 6th his ship, the frigate *Gloucester*, was wrecked off the Norfolk coast, and nearly 150 members of his entourage were drowned, including several of his musicians; he finally arrived back at Whitehall on 27 May.[23]

It is not immediately clear whether Purcell's ode was performed in March or May, but the lack of references to the *Gloucester* debacle in the text is perhaps an argument for assigning it to his first return from Scotland. The poet picks his way carefully through the delicate political situation, praising James's courage and his loyalty to the crown ('York the obedient, grateful, just, | Courageous, punctual, mindful of his trust.'), alluding to threats of plot and rebellion, and defiantly affirming his status as heir-apparent: 'And now ev'ry tongue shall make open confession | That York, royal York, is the next in succession.'

In some respects, 'What shall be done' is a less advanced and developed work than 'Swifter, Isis'—the key scheme is less varied and there are no ground basses—though it is by no means inferior as music. It has, for instance, one of the young Purcell's finest overtures, with an extended and beautifully worked-out fugue in hornpipe rhythm, full of teasing cross-rhythms. Like 'Swifter, Isis', 'What shall be done' uses two treble recorders as well as strings; the French players may have left England shortly after it was performed, for wind instruments are not called for again in a court ode until 'Ye tuneful Muses' of 1686. A declamatory bass solo, 'Mighty Charles, though joined with thee', again acts as a kind of pivot, around which are grouped vocal minuets, and the minuet-like ritornello for recorders and strings that ends the first choral section recurs just before the last chorus. Also, the fourth number, the solo and chorus 'All the grandeur he possesses', set to a haunting bourée-like tune (which Purcell later used in the suite for *The Gordion Knot Unty'd* Z597/2), is balanced at the end of the work by the delightful vocal gavotte 'May all factious troubles cease'.

Purcell's next three court odes can be passed over fairly rapidly, for they are similar in style to 'Swifter, Isis' and 'What shall be done', and are largely inferior to them in quality. The first, 'The

[23] Luttrell, *Brief Relation*, i. 171, 177, 181, 184–5, 189; see H. Love, 'The Wreck of the Gloucester', *MT* 125 (1984), 194–5; Holman, *Four and Twenty Fiddlers*, 418–19.

summer's absence unconcerned we bear' Z337, was written 'for his Majesty at his return from New Market October the 21–1682', according to the heading in R.M. 20.h.8, while the second, 'From hardy climes and dangerous toils of war', was 'perform'd to Prince George [of Denmark] upon his Marriage w(i)th the Lady Ann' (the future queen) on 28 July 1683. The third, 'Fly, bold rebellion' Z324, is labelled 'The Welcome Song perform'd to his Majesty in the year 1683', but the text is concerned with the Rye House plot, which came to light that spring. Thus, it may have been performed on 9 September, the day appointed to celebrate deliverance from the plot, when the king was in Winchester, or later that month, when he eventually returned to Whitehall.[24]

These works mark the beginning of a trend to introduce more counterpoint into the ode. As we have seen, the same thing happened in the anthem at about the same time, and for the same reason: by casting some of his choruses as fugal movements, and by writing solo passages for several solo voices in an idiom drawn from the verse sections of symphony anthems, Purcell and Blow were able introduce greater variety into the genre, making it less dependent on dance patterns. Both 'The summer's absence' and 'Fly, bold rebellion' have anthem-like opening vocal sections, the former for ATB soloists, the latter for AATBB, and both end with fugal passages; in 'Fly, bold rebellion' there is a remarkable contrapuntal ensemble in minuet time for seven solo voices, which leads into the final six-part chorus. Another sign of the times is Purcell's use of fugal writing in the first strain as well as the second of the fine symphony to 'The summer's absence'. He used the same device in a number of later odes, including the 1683 St Cecilia ode 'Welcome to all the pleasures' Z339, as did Blow in 'Begin the song', the 1684 St Cecilia ode.

The odes of 1682–3 also illustrate the development of Purcell's ground-bass writing. In 'Swifter, Isis' there is a true ground bass, 'Hark! just now my list'ning ears', constructed over an unvaried duple-time C major fanfare motif, as well as the duet 'The king, whose presence like the spring'. The latter is an example of the type I have christened 'pseudo-ground' because, though its patterns of flowing quavers in the bass make it sound like a ground bass, and recall works in similar style that are real ground basses (such as 'Oft she visits this lone mountain' in *Dido and Aeneas*

[24] Luttrell, *Brief Relation*, i. 279, 281.

Z626/25), the piece is actually a binary song in which the bass pattern only fits the first phrase of the vocal line, and changes immediately the voice moves on to the second phrase. Real ground basses, by contrast, fit a number of the vocal phrases, and only depart from the pattern if at all to effect key changes. 'The summer's absence' has a true ground, the alto solo and chorus 'And when late from your throne', set to an unvaried major-key *passacaglia* bass, as well as a pseudo-ground, the alto solo 'These had by their ill usage drove', also strikingly similar to 'Oft she visits'.

In the tenor solo 'The sparrow and the gentle dove', the ground in 'From hardy climes', Purcell brings elements of the two types together. It is a true ground, unvaried until Purcell allows it to modulate during the final ritornello (a striking touch), but the melody still falls into repeated binary phrases, generating a fascinating tension between its harmonic implications and those of the bass (Ex. 5.2). The piece is also the first example in Purcell of a type tried out by Blow in the duet 'All due, great prince' (from the 1683 New Year ode 'Dread Sir, Father Janus', Add. MS

Ex. 5.2. 'The sparrow and the gentle dove' from 'From hardy climes and dangerous toils of war' Z235/8, bars 245–54

Ex. 5.2. *cont.*

33287, fos. 125ʳ–130ʳ) in which the ground is arpeggiated so that it effectively consists of a two-part dialogue between a simple bass and an interlocking tenor. Purcell evidently liked the effect, for he used it again in 'Be welcome then, great Sir' in 'Fly, bold rebellion', and many times in later odes. Both of these ground basses derive much of their effect from their ravishing string ritornelli, whose rich harmonies give the movements a new lease of life just as the ear is beginning to tire of them.

By the time 'Fly, bold rebellion' was performed Purcell must have been engaged on a new enterprise, the composition of a St Cecilia ode. The origin of the English custom of celebrating St Cecilia's Day, 22 November, with a concert featuring the performance of a specially composed ode is shrouded in mystery. In fact, virtually everything known about the first datable ode, Purcell's 'Welcome to all the pleasures', comes from the title-page of the score, published by John Playford in 1684:

> A Musical Entertainment PERFORM'D ON NOVEMBER XXII. 1683. IT BEING THE Festival of St. Cecilia, a great Patroness of Music; WHOSE MEMORY is ANNUALLY Honour'd by a Public Feast made on that Day by the MASTERS and LOVERS of Music, as well in England as in Foreign Parts.

It is dedicated to 'THE GENTLEMEN OF THE Musical Society', though it is not clear whether this body was newly formed or had been in existence for some time. Four stewards are named 'For the YEAR ensuing'. William Bridgeman was Clerk of the Privy Council, while Gilbert Dolben was a young lawyer; the others,

Nicholas Staggins and Francis Forcer, were musicians, so it looks as if the enterprise was jointly run by amateurs and professionals. We do not know where the 1683 performance took place, but the 1684 celebrations were held in Stationers' Hall, which was the venue for succeeding years.[25]

The situation is complicated by the existence of two small-scale odes that have also been allocated to 22 November 1683. 'Laudate Ceciliam' Z329 was certainly written in that year: in R.M. 20.h.8 Purcell headed it 'A Latine Song made upon St Cecilia, whoes day is comme[mo]rated yearly by all Musitians made in the year 1683'. But we do not know whether it was performed with 'Welcome to all the pleasures', or in some more private milieu—such as the queen's Catholic chapel. 'Raise, raise the voice' Z334 is not dated in any early source, and may not be a St Cecilia ode: the text does not mention the saint, and is concerned with praising Apollo on 'sacred Music's holy day'. In fact, the work has more in common with Purcell's symphony songs than with his odes, and like them may have been written for some informal musical entertainment at court. There is certainly no vacancy for a St Cecilia ode in the years after 1683. The 1684 ode, 'Begin the song', was set by Blow, while William Turner and Isaac Blackwell produced the 1685 and 1686 odes, though their settings do not survive; the 1687 ode, Dryden's 'From harmony, from heavenly harmony', was set by Giovanni Battista Draghi (see below).[26]

'Welcome to all the pleasures' is by far the best-known of Purcell's early odes, but it does not entirely live up to its reputation. The main problem is that the most substantial numbers—the symphony, the opening anthem-like sequence of verse and chorus passages, and the fine alto solo on a ground 'Here the deities approve'—all come in the first half. The rest is set entirely in minuet-like triple time, except for the strange four-bar continuo passage before the tenor solo 'Beauty, thou scene of love', which is obviously inserted to bring the music back from A minor to the home key E minor, but actually serves mainly to draw the listener's attention to the need at that point for a substantial duple-time movement to balance 'Here the deities approve'. Incidentally,

[25] The only published survey of the subject is still W. H. Husk, *An Account of the Musical Celebrations on St. Cecilia's Day* (London, 1857).

[26] J. Blow, *Begin the Song*, ed. H. W. Shaw (London, 1950); Husk, *An Account*, 16–20; I am grateful to Bruce Wood for informing me that there is a copy of the printed text of Blackwell's ode at GB-Cu.

this last, another example of the arpeggiated 'dialogue' type of ground, exists in a beautiful keyboard arrangement, published by the composer in *The Second Part of Musick's Hand-Maid* (1689), with *style brisé* harmonizations of the ground that give us a vivid idea of how Purcell might have realized the continuo.

'Raise, raise the voice' is arguably a more successful work. The anonymous text is similar in its tone and phraseology to the text of 'Welcome to all the pleasures', and may therefore be by the same author, the minor poet and musician Christopher Fishburn; Zimmerman actually ascribed it to '(?Christopher) Fishburn', but without revealing the source of his information.[27] The settings, however, are rather different in style. Purcell seems to have written 'Raise, raise the voice' just for three solo voices (the solo and chorus indications in *Works II*, 10 are editorial), but he filled out the texture by giving the violins independent parts in the vocal sections. In earlier odes the strings had been confined largely to alternating with the voices, or to doubling them in tutti sections, though some choruses have an obbligato violin part, and both violins have independent parts in the last chorus of 'Welcome to all the pleasures'. In 'Raise, raise the voice' they have largely independent parts in the first vocal section, creating a bright SSSTB texture, and they join the soprano in the ground bass 'Mark how readily each pliant string', gently echoing the voice as well as contributing a beautiful and idiomatic final ritornello (Ex. 5.3). English composers had been slower than their counterparts on the Continent to develop sophisticated ways of combining voices and violins, and movements of this sort were still a novelty to them.

In this respect, 'Laudate Ceciliam' is less advanced than 'Raise, raise the voice'—which may mean that the latter was written a year or two after 1683. Purcell never combines the violins with the ATB voices, even though this means that they apparently do not play at the end of the work, though it is conceivable that they would have doubled the vocal lines in some way in the ensemble passages. However, both works show how rapidly Purcell's sense of structure was maturing in the early 1680s. The latter consists essentially of four large sections: (1) a complex consisting of the symphony, the opening bass solo and tutti section, and a repeat of the first strain of the symphony, all in duple time; (2) a matching complex consisting of a soprano solo, a tutti section, and a ritor-

[27] Zimmerman, *Purcell: Catalogue*, 157.

Ex. 5.3. 'Mark how readily each pliant string' from 'Raise, raise the voice' Z334/7, bars 203–10

nello, largely in minuet rhythm; (3) a ground, consisting of a soprano solo, a tutti passage, and a ritornello; and (4) a concluding tutti in minuet rhythm. Listeners easily perceive that sections 1 and 3 and 2 and 4 balance one another, but they are also aware of a bipartite division between sections 1 and 2, which are in D minor, and sections 3 and 4, in D major.

'Laudate Ceciliam' also divides into four main sections, though the relationship between them is more complex. The work has an unusually strong feeling of coherence, partly because all the sections end in the home key, C major, and partly because Purcell uses only two types of rhythmic movement in the vocal sections: duple time, predominantly in patterns of quavers and semiquavers

(D), and a graceful triple time (T), originally written in void notation, probably to indicate a leisurely tempo. The first section, the symphony (D, T), is repeated complete in the middle of the work (a procedure commonly found in anthems at the time), and the first vocal complex (T, D, T) balances the second (D, T, D, T), the more so since the two Ts of the former, and the second of the latter are a refrain to the words 'Laudate Ceciliam'. This structure seems to have been determined by Purcell rather than the anonymous author of the words, and was probably inspired by mid-seventeenth-century Italian motets, which frequently have recurring refrains of this sort. Indeed, 'Laudate Ceciliam', with its sweet and sensuous passages of triple time and its absence of dance-based structures, is one of Purcell's most Italianate works.

Purcell's remaining welcome odes can be considered together. 'From those serene and rapturous joys' Z326 is a setting of a poem by Thomas Flatman, who entitled it 'On the King's return to White-hall, after his Summer's Progress, 1684'; that year the king reached Whitehall from Winchester on 25 September.[28] In R.M. 20.h.8 'Why, why are all the Muses mute?' is entitled 'Welcome Song 1685 being the first Song performed to King James the 2d.', while the copies of 'Ye tuneful Muses, raise your heads' Z344 and 'Sound the trumpet, beat the drum' Z335 are entitled 'Welcome Song 1686' and 'Welcome Song 1687'.

Taken literally, these headings mean that the works were first performed on or soon after 6 October 1685, 1 October 1686, and 11 October 1687, the days James reached Whitehall respectively from Windsor, Windsor again, and Oxford after his summer progresses.[29] But the lack of surviving odes for James II's birthday has led writers to suggest that they were actually performed on 14 October, and it must be said that Pepys's description of the welcome ode performed on 10 October 1685—'Wherin the fraequent Returnes of the Words, Arms, Beauty, Triumph, Love, Progeny, Peace, Dominion, Glory &c. had apparently cost our Poët-Prophet more paine to finde Rhimes then Reasons'[30]—does not correspond to the text of 'Why, why are all the Muses mute?'.

Of the four, 'Why, why are all the Muses mute?' is by far the best. Perhaps responding to the challenge of pleasing a new monarch, and certainly inspired by a text that offered him plenty

[28] G. Saintsbury (ed.), *Minor Poets of the Caroline Period* (London, 1921), iii. 377–8; Luttrell, *Brief Relation*, i. 316.

[29] Ibid. i. 359, 385, 415.

[30] Pepys, *Letters and the Second Diary*, 171.

of opportunities for vivid word-painting, Purcell produced his longest, most varied, and most dramatic ode to date. The drama starts in the first bar: a countertenor accompanied only by continuo commands the silent instruments to awake; the strings do not play until after the succeeding chorus, when they embark on a symphony in the normal two-section French pattern. The device was not new—Purcell did something similar at the beginning of the anthem 'My heart is fixed, O God' Z29 (?1682–3)—though nothing could have prepared James II and his court for the operatic atmosphere of Purcell's ode.

There are several other memorable moments later in the work. One is at the end of the soprano duet 'So Jove, scarce settled in his sky', at a point where the poet has been alluding in lurid terms to Monmouth's rebellion. Anticipating the next couplet, 'Caesar, for milder virtues honour'd more, | For his goodness lov'd than dreaded for his pow'r', Purcell turns from G minor to B flat major, and powerfully evokes the era of peace supposedly ushered in by the new reign with an exquisite minuet in rondeau form (also used in *The Gordion Knot Unty'd* suite Z597/3)—another movement in an ode that cries out to be danced. Best of all is the conclusion of the ode, where Purcell matches the last sentence—'His fame shall endure till all things decay, | His fame and the world together shall die, | Shall vanish together away.'—with a matchless dying fall (Ex. 5.4).

'Sound the trumpet, beat the drum' is not such an arresting work, though it was the only welcome ode to retain its popularity in the 1690s, and according to Charles Burney the duet on a ground 'Let Caesar and Urania live' 'continued so long in favour, not only while those sovereigns [William and Mary] jointly wielded the sceptre, but even when George II. had lost his royal consort, and there ceased to be a Queen, or Urania, for whom to offer up prayers, that Dr. Green, and afterwards Dr. Boyce, used frequently to introduce it into their own and the laureate's new odes'.[31] The piece is the prototype for a number of energetic and captivating countertenor duets on running ground basses in Purcell's later odes—such as 'Sound the trumpet' in 'Come, ye sons of art, away' Z323/3.

Otherwise, 'Sound the trumpet, beat the drum' is mainly of interest for the new ways in which instruments are combined with

[31] Burney, *General History*, ii. 393.

Ex. 5.4. 'O how blest is the isle' from 'Why, why are all the Muses mute?' Z343/13, bars 579–601

Ex. 5.4. *cont.*

591
all things de - cay, till all things de - cay, His
all things de - cay, till all things de - cay, His
- cay, en - dure till all things de - cay, His
- cay, till all things de - cay, de - cay, His
595
fame and the world to - ge - ther shall die, shall
fame and the world to - ge - ther shall die, shall
fame and the world to - ge - ther shall die, shall
fame and the world to - ge - ther shall die, shall
599
1
2
van - ish to - ge - ther a - way. His way.
van - ish to - ge - ther a - way. His way.
van - ish to - ge - ther a - way. His way.
van - ish to - ge - ther a - way. His way.

voices. All the choruses have passages in which the strings play brief interludes between the vocal phrases, and 'With plenty surrounding' has a good deal of independent writing for the upper strings, often making seven real parts. 'While Caesar, like the morning star' is the first of a long line of solos in Purcell's odes in which a solo bass is accompanied throughout by four-part strings, though there are bass solos accompanied by two solo violins in 'From those serene and rapturous joys' and 'Why, why are all the Muses mute?', and the type seems to have had its origin in 'Music's the cordial of a troubled breast' from Blow's 'Begin the song'. English composers tended to reserve string accompaniment for bass voices until well into the 1690s, perhaps because they thought higher voices would be drowned by anything more than continuo instruments. Incidentally, the presence of passages for two violins and continuo in these odes is a sign that an orchestra was used to play the four-part string passages. Draghi specifically allocates the former to single instruments in 'From harmony, from heavenly harmony', contrasting them with a five-part passage marked 'here enter all the viol[ins] and other instruments'.[32]

The best movement in 'Ye tuneful Muses' is 'To music's softer but yet kind and pleasing melody', a beautiful AAB trio with two recorders, though the original audience would probably have remembered the alto solo 'Be lively, then and gay'. The latter is one of several pieces that Purcell constructed over a popular tune, played in the bass part. He seems to have used this strange device to deliver a hidden message to his audience, witness Hawkins's story about 'May her blest example chase' in 'Love's goddess sure was blind', based on 'Cold and raw' (see Ch. 1). 'Be lively, then and gay' uses the tune 'Hey, boys, up go we', which was associated at the time with a ballad Thomas D'Urfey wrote to satirize the pretensions of the Whigs in the Exclusion Crisis of 1679–81, drawing on an anti-Puritan song by Francis Quarles from the early 1640s.[33] So the message on this occasion was not all that hidden: James II was being reassured that his opponents would eventually be defeated, just as the Puritans had been in 1660, and the Whigs had been in 1681–2 (Ex. 5.5). Interestingly, a third

[32] Holman, *Four and Twenty Fiddlers*, 427–8; see also 374–5 for an argument that the passages for two violins and continuo in Blow's *Venus and Adonis* imply the use of an orchestra for the four-part writing.

[33] W. Chappell, *The Ballad Literature and Popular Music of the Olden Time* (London, 1859; repr. 1965), ii. 425–9; C. M. Simpson, *The British Broadside Ballad and its Music* (New Brunswick, NJ, 1966), 304–8.

Ex. 5.5. 'Be lively, then and gay' from 'Ye tuneful Muses' Z344/5, bars 136–59

piece of the same sort, the jig in the suite for *The Gordion Knot Unty'd* Z597/5, has 'Lilliburlero' in the bass, the ballad said to have 'sung a deluded prince out of three kingdoms'.[34] *The Gordion Knot Unty'd* is a compilation suite, partly put together from the ritornelli of court odes of the 1680s, and it may be that its jig comes from a lost ode of 1689–90—perhaps even from one that celebrated the Glorious Revolution. Purcell's colleagues must have smiled at his use of the same contrapuntal device to convey such conflicting political messages.

The first performance of Draghi's 'From harmony, from heavenly harmony' on 22 November 1687, just over a month after 'Sound the trumpet, beat the drum' was first performed at court, marks a turning-point in the history of the English ode. It was the first setting of an ode by a major poet, John Dryden; it was the first composed by a immigrant musician; it was the first to include parts for trumpets and drums; and the first to use the Italianate five-part string layout of two violins, two violas, and bass (see Ch. 3). Moreover, it was the first major choral work to be written in England in the contemporary Italian style. It is on a much larger scale than its predecessors; it opens with a descriptive Italianate prelude instead of the traditional French overture; it has massive choruses based for the first time on Italianate counterpoint rather than French dance patterns; and its solos are more florid and extended than anything written up to that time in England.

'From harmony, from heavenly harmony' evidently caused something of a sensation among English musicians. Five early scores of it survive, and its influence can be heard immediately in Purcell's symphony anthem 'Behold, I bring you glad tidings' Z2, written for Christmas Day 1687, as well as in his subsequent odes and anthems. Anyone who has heard the work would find it hard to agree with Franklin Zimmerman that Draghi's 'poetic sensibilities and musical powers were not such as to enable him to take full and imaginative advantage of the opportunity Dryden had provided'.[35] It has its rough edges, but it also has a vividness and grandeur that was matched by Purcell only in his most ambitious odes, and it is no exaggeration to assert that the younger composer could not have written 'Hail, bright Cecilia' without studying it.

[34] Chappell, *Ballad Literature*, ii. 568–74; Simpson, *British Broadside Ballad*, 449–55.

[35] Zimmerman, *Purcell: Life and Times*, 147; see also E. Brennecke, 'Dryden's Odes and Draghi's Music', *Proceedings of the Modern Language Association of America*, 49 (1934), 1–36.

Draghi's influence is particularly strong in the odes Purcell wrote in 1689–90. The earliest is 'Now does the glorious day appear' Z332, a setting of a text by the new Poet Laureate Thomas Shadwell, and the first of six odes for Queen Mary's birthday; it was performed on 30 April 1689. Next came an occasional ode, 'Celestial music did the gods inspire' Z322, headed in R.M. 20.h.8 'A Song that was perform'd at Mr Maidwells a school master on the 5th of August 1689 The words by one of his scholars.' The Revd Louis Maidwell ran a school at his house in King Street, Westminster, and the ode was presumably performed there, perhaps with the help of some of Purcell's court colleagues. The text, a surprisingly accomplished piece of work for a child, is in praise of music; not surprisingly, it echoes 'From harmony, from heavenly harmony' in one or two places.

Purcell's next ode was also written for a special occasion outside the court. 'Of old when heroes thought it base' Z333, The Yorkshire Feast Song, is a setting of a text by Thomas D'Urfey, who eventually published it under the title 'An ODE on the Assembly of the Nobility and Gentry of the City and County of York, at the Anniversary Feast, March the 27th. 1690. Set to Musick by Mr. Henry Purcell. One of the finest Compositions he ever made, and cost 100*l.* the performing.'[36] Contemporary newspaper reports confirm that the ode was indeed performed on that day in Merchant Taylors' Hall—there is mention of 'a very splendid Entertainment of all sorts of Vocal and Instrumental Musick'—but also reveal that the event had been postponed, from 14 February.[37] Purcell's second ode for Queen Mary's birthday, 'Arise, my Muse' Z320, another setting of D'Urfey, was performed just over a month later, on 30 April 1690.

It seems to have taken some time for Purcell to come to terms with all the novel features of 'From harmony, from heavenly harmony'. In 'Now does the glorious day appear' its influence is mainly confined to the choral and orchestral sections. Purcell uses Draghi's five-part string-writing, and his ode begins with a fine Italianate symphony, probably the first by an English composer that begins with a fast contrapuntal passage based on a fanfare motif rather than the normal passage in dotted rhythms. The first chorus is also Italianate in style. It is a splendidly vigorous

[36] T. D'Urfey (ed.), *Songs Compleat, Pleasant and Divertive* (London, 1719; repr. 1872), i. 114–16.

[37] M. Tilmouth, 'Calendar', 9; Westrup, *Purcell*, 65–6.

contrapuntal movement based on two ideas, the second of which generates immense energy by the simple device of speeding up the harmonic rhythm (Ex. 5.6). Otherwise, the work is oddly conventional, even conservative in style, with two pseudo-grounds of the type favoured by Purcell in the early 1680s, and a ground-bass movement, 'By beauteous softness mix'd with majesty', with a melodic line cast in a binary pattern—another early feature.

'Celestial music' has received little attention, perhaps because it starts with the symphony of the 1685 coronation anthem 'My heart is inditing' Z30, which may have given the impression that the ode was put together in haste. In fact, it is a stronger work than 'Now does the glorious day appear', and in it Purcell extended Draghi's idiom to some of the solo movements. For instance, Purcell seems to have taken Draghi's setting of 'The soft complaining flute' as the starting-point for 'Her charming strains expel tormenting care'. Both movements are in the same key, C minor, both are modulating ground basses, and both are scored for countertenor with two recorders and continuo. In particular, Purcell followed Draghi in using the recorders in a new way: as well as providing the customary final ritornello, they clothe the voice in rich harmony and provide interludes between the vocal phrases. This beautiful movement shows that Purcell had finally come to terms with the sophisticated and varied ways Italian composers had devised for combining voices and obbligato instruments. There are a number of other points of interest in 'Celestial music'. The fine countertenor solo 'When Orpheus sang all nature did rejoice' is the first in Purcell's odes in which a high voice is accompanied by four-part strings, while in the final chorus the phrases of a minuet are separated by sparkling Italianate ritornelli—an early example of the French and the Italian styles reconciled in a single movement.

'Celestial music' only uses recorders and strings, presumably because Maidwell's school did not have the resources or the space for a large band, and it was not until the Yorkshire Feast Song that Purcell came to terms with one of the most significant innovations of Draghi's ode, the introduction of trumpets into the orchestra. The trumpet began to appear in concerted instrumental music on the Continent around 1650; hitherto it had played only a peripheral role in art music, and had been used mainly as a fanfare instrument, though Schütz, Praetorius, and other German composers had experimented with trumpet bands in church music.

Ex. 5.6. 'Now does the glorious day appear' Z332/2, bars 82–94

Ex. 5.6. *cont.*

87
- rious day ap - pear, ap - pear,
Now does the glo - - - -
- pear, ap - pear,
Now does the
Now does the glo - -
89
[tr]
[tr]
Now does the glo - - - - rious day ap -
tr
- - - - - - - rious day ap -
glo - - - - - - rious day ap -
- - - - - - - rious day ap -

Ex. 5.6. *cont.*

91
- pear,
- pear,
The migh-tiest day of all the
- pear, The migh-tiest day of all the year, of all the
- pear,
93
[tr]
The migh-tiest day of all the year,
[tr]
year, of all the year, The migh-tiest, migh-tiest day of all,
year, The migh-tiest, migh - tiest day,
The migh-tiest day of all the year, The migh-tiest day of all

Maurizio Cazzati's *Suonate à due, trè, quattro, e cinque, con alcune per tromba* Op. 35 (Bologna, 1665) contains the earliest datable pieces for solo trumpet and strings, although Johann Heinrich Schmelzer had already published a sonata for two trumpets and strings in his *Sacroprofanus concentus musicus* (Nuremberg, 1662), and there is a 'Sonata a 5. 2 Violini e 2 Trombette con Fagotto' by Vincenzo Albrici at S-Uu, IMhs 1:3 that seems to have been written between 1652 and 1654, when Albrici was director of the Italian musicians at the Swedish court.[38] The latter is particularly important because Albrici was the leader of the consort of Italian singers who came to England in 1663 and were given posts at court; Draghi joined it a year or two later.[39] Samuel Pepys was probably referring to Albrici or Draghi when he wrote in his diary for 2 November 1666 that 'the King's Italian here [at Whitehall] is about setting three parts for Trumpets and shall teach some of them [the royal trumpeters] to sound them, and believes they will [be] admirable Musique'.[40]

Draghi's ode was not the first English work to use trumpets with other instruments and/or voices. A chorus in Locke's music for *Psyche* (1675) was accompanied by 'Trumpets, Kettle-Drums, Flutes and Warlike Musick'; a lost instrumental piece for trumpets and orchestra was played during the celebrations for James II's coronation in 1685; and portions of a 'Concerto di trombe a tre trombette con violini e flauti' were published by Nicola Matteis in 1685, and in a revised form in 1687.[41] Peter Downey has argued that the string-writing in fanfare style that appears in 'Ye tuneful Muses' and 'Sound the trumpet, beat the drum' should be construed as evidence of missing trumpet parts rather than just imitations of trumpets. But Draghi's ode is the first surviving English concerted work with unambiguous indications for

[38] D. Smithers, *The Music and History of the Baroque Trumpet before 1721* (London, 1973), 95–8; Holman, *Four and Twenty Fiddlers*, 428; J. H. Schmelzer, *Sacro-profanus concentus musicus (1662)*, ed. E. Schenk (DTÖ 111/112; Graz and Vienna, 1965), no. 1; A. Schering (ed.), *Geschichte der Musik in Beispielen* (Leipzig, 1931), no. 214.

[39] For this group, see in particular Westrup, 'Foreign Musicians in Stuart England'; Mabbett, 'Italian Musicians in Restoration England'.

[40] Pepys, *Diary*, vii. 352.

[41] Holman, *Four and Twenty Fiddlers*, 328–30, 347–8, 428–9; N. Matteis, *Concerto in C*, ed. P. Holman (London, 1982); P. Downey, 'What Samuel Pepys Heard on 3 February 1661: English Trumpet Style under the Later Stuart Monarchs', *EM* 18 (1990), 417–28; A. Pinnock and B. Wood, 'A Counterblast on English Trumpets', *EM* 19 (1991), 436–43, and my response, ibid. 443.

trumpets, and it provided the model for the trumpet-writing in Purcell's subsequent odes.

In fact, the trumpet-writing in The Yorkshire Feast Song is already more advanced than that in 'From harmony, from heavenly harmony' in one crucial respect. In the surviving scores of Draghi's ode the trumpet parts are on the violin staves, which means that it is sometimes not clear exactly what they should play. In The Yorkshire Feast Song they are given separate staves, and have independent parts in several movements. The two-strain symphony, though short and relatively unsophisticated, is probably the earliest example of an Italianate trumpet sonata by an Englishman.

The other novel feature of the instrumental writing in the work is the heading 'For Violins & Hoboys' in the extended prelude to the alto solo 'The pale and the purple rose', a fine pseudo-ground. The oboe parts, like Draghi's trumpet parts, are written on the string staves, so it is not always clear exactly what notes they should play, or whether there should also be a tenor oboe and a bassoon. But the example of some of the movements in *Dioclesian* Z627 (June 1690), with their antiphonal passages for a double-reed quartet and four-part strings, suggests that there should. Indeed, 'The pale and the purple rose' probably served as a model for them.

In truth, The Yorkshire Feast Song is more remarkable for its scoring than for the quality of its music, though it has one of Purcell's loveliest ground basses, 'So when the glitt'ring queen of night', a tenor solo richly accompanied throughout by four-part strings. The work was the prototype for most of Purcell's late odes in that it uses a complete Baroque orchestra, with trumpets, oboes, recorders, strings, and continuo, and that the trumpets are prominently deployed, giving it a triumphant, martial cast. This reflected the increasingly bellicose flavour of the texts Purcell set, which in turn reflected the bellicose temper of English society in the 1690s, taken up as it was with cheering on William III's annual expeditions against the French.

It also doubtless reflected the king's own musical taste, so far as it can be identified. The younger John Banister wrote in his oboe treatise *The Sprightly Companion* (London, 1695) that, 'besides its Inimitable charming Sweetness of Sound (when well play'd upon)', the instrument 'is also Majestical and Stately, and not much Inferiour to the Trumpet; and for that reason the greatest

Heroes of the Age (who sometimes despise Strung-Instruments) are infinitely pleased with This for its brave and sprightly Tone'. William, the Dutch Hero of the Age, did much to promote oboe bands and trumpeters in the army and at court, to the detriment of the 'strung instruments', the Twenty-four Violins.[42]

'Arise, my Muse' is a work in this mould, with a fine martial symphony for two trumpets and five-part strings (reused in *King Arthur* Z628/4), and a florid ritornello for two trumpets and two oboes at the end of the ground bass 'Hail, gracious Gloriana, hail!'. But the ode has a puzzling feature. D'Urfey's text, published in 1698, has two more verses than the version set by Purcell, and, as Bruce Wood has pointed out, the setting may be incomplete, for it finishes abruptly without bringing back the trumpets and oboes; it is possible that the work survives incomplete, though it is perhaps more likely that Purcell failed to complete it for some reason.[43] Nevertheless, this last section contains some arresting music. It begins as a declamatory D minor solo, 'But ah, I see Eusebia drown'd in tears', in which a countertenor accompanied by two recorders vividly describes the queen's anguish as her husband departs for the wars. From time to time a bass accompanied by two solo violins bursts in with a bouncy D major response in jig rhythm 'But Glory cries, "Go on, illustrious man"', which is eventually taken up by the chorus and orchestra (Ex. 5.7). It is of note that this powerful dramatic situation seems to have been created by Purcell rather than D'Urfey: the two passages are just placed consecutively in the published version of the ode, with no sign that the poet wanted them to be interspersed as a dialogue.[44]

'Welcome, welcome' glorious morn' Z338, Queen Mary's 1691 birthday ode, is a fine vigorous work, full of the pompous trumpetings so fashionable at the time; its Italianate symphony is the first in which the oboes as well as the trumpets are given separate staves and independent parts. However, it is mainly of note for a sequence that provides something akin to the contrast between recitative and aria in contemporary Italian concerted works. The pseudo-ground basses 'To lofty strains her tuneful lyre she strung' and 'I see the round years successively move' are preceded

[42] On this point, see Holman, *Four and Twenty Fiddlers*, 432–4.

[43] H. Purcell, *Birthday Odes for Queen Mary, Part 1*, ed. B. Wood, *Works II*, 11 (London, 1993), pp. ix–x.

[44] Ibid. xx–xxi.

Ex. 5.7. 'But ah, I see Eusebia drown'd in tears' from 'Arise, my Muse' Z320/9, bars 383–403

Ex. 5.7. *cont.*

393
no, Fate must not let him go, Fate must not, no,
397
Violin 1 [solo]
Violin 2 [solo]
must not let him go." But Glo - ry cries, "Go on, go
400
on, Go on, great Prince, go on, Go on, great Prince, go on."

by recitative-like declamatory solos for the two soloists, respectively tenor (but allocated to a bass in *Works II*, 11) and soprano. The 'recitatives' are Italianate enough, with some brilliant semiquaver passage-work, though the 'arias' are still stubbornly English in form. They are both essentially binary movements, and the words and music of 'To lofty strains' are eventually taken up by the chorus, as in many numbers in the earlier odes.

Purcell's ode for Queen Mary's birthday in 1692, 'Love's goddess sure was blind' Z331, is also a fine work, though it has quite a different character from 'Welcome, welcome glorious morn'. Sedley's text is refreshingly intimate and reflective in tone, and this is reflected in the music by the absence of wind instruments, and by the use of predominantly minor keys. Thus, the flashy Italianate idioms developed by Purcell in the previous few odes are conspicuous by their absence. There is, for instance, a fine G minor French overture (reused in the play *The Rival Sisters* Z609/1) instead of a fanfare-like trumpet sonata. Nevertheless, the work has an outstanding number that demonstrates the extent to which Purcell had assimilated Italian harmonic thinking. The duet 'Sweetness of nature and true wit', for two countertenors with two solo violins (wrongly allocated to recorders in *Works I*, 24), is a flowing, melodious piece in slow triple time, similar to some of the Italian songs in Pignani's 1679 collection (see Ch. 2). Its structure is determined not by repetition—as in ground basses, binary dance songs, or da capo arias—but just by a logical sequence of modulations, moving gently but purposefully through a series of related keys: G minor, B flat major, C minor, F major, D minor, F major, and G minor.

'Hail, bright Cecilia', a setting of a poem by the Revd Nicholas Brady (best known today for the collection of metrical psalms he produced with Nahum Tate), is Purcell's largest and grandest choral work. It was also his most popular, to judge from the number of surviving sources. They include GB-Ob, MS Mus. C. 26, fos. 21^{r}–69^{v}, a largely autograph score apparently used for at least two early performances; GB-Cfm, Mu. MS 119, pp. 1–55, a score that the copyist, William Croft, labelled 'for the 22 of Frebruary 1695[/6]', possibly recording a performance on that date; and the surviving portions of two sets of parts, GB-Ob, Tenbury MS 1309 and MS Mus. C. 27, fos. 3^{r}–25^{r}. The latter do not seem to derive from Purcell's own performances—Tenbury MS 1309 probably dates from around 1700, and MS Mus. C. 27 is much later—but they are virtually the only near-contemporary performing material we have for Purcell's major concerted works, and Tenbury MS 1309 contains important information about the disposition of the instruments. It shows, for instance, that the oboes should double the trumpets rather than the violins when they do not have their own parts—a point confirmed by Mu. MS 119.

There is also important information about the solo singers in MS Mus. C. 26. A number of early sources of court odes give the names of the soloists of particular numbers, but the autograph of 'Hail, bright Cecilia' has the most complete information, with some names crossed out and replaced by others—apparently to take account of the casts of later performances. It shows that Purcell used at least thirteen solo singers, including a woman, Mrs Ayliff, his leading theatre soprano. This may have been an innovation, for earlier odes are likely to have used all-male casts, with boys from the Chapel Royal. Indeed, the earliest score of Draghi's ode, West Sussex Record Office, Cap. VI/1/1, fos. 24^{r}–63^{r}, allocates solo soprano passages to 'one of the Boys'.

In the autograph of 'Hail, bright Cecilia' Purcell gave the spectacular countertenor solo ''Tis Nature's voice' to John Pate, but Peter Motteux wrote in *The Gentleman's Journal* for November 1692 that the ode 'was admirably set to Music by Mr. Henry Purcell, and perform'd twice with universal applause, particularly the second Stanza, which was sung with incredible Graces by Mr. Purcell himself'.[45] This has given rise to the notion that Purcell was a countertenor, and sang the solo at the first performance, but he is known to have been a bass (see Ch. 1), and the phrase is more likely to mean that the graces were composed rather than sung by Purcell, for the piece has most unusual and elaborate written-out ornamentation. It is true that Thomas Cross described it 'sung by himself at St Caecilia's Feast' when he published it in, probably, 1693, but Cross could easily have been misled by Motteux's ambiguous sentence, as I believe others have been in our own time.

'Hail, bright Cecilia' contains some superb music. ''Tis Nature's voice' is Purcell's most extravagant and imaginative declamatory solo, while the three ground basses, 'Hark each tree', 'Wondrous machine', and 'In vain the am'rous flute', are unsurpassed examples of their particular genre, though they are quite different from

[45] Purcell, *Ode on St Cecilia's Day 1692*, ed. P. Dennison, *Works II*, 8 (London, 1978), p. ix.

each other. Nevertheless, the listener cannot fail to sense that the work is greater than the sum of its parts; this is partly because it is laid out on an unprecedented scale, and partly because Purcell achieves a fine balance between unity and diversity in matters of structure. The Italianate symphony, for instance, is one of the longest works of its type written anywhere in Europe up to that time—it plays for about ten minutes with all repeats. Yet the ear does not tire of the festive trumpetings because the D major sections—a fanfare-like passage (A), a canzona on two subjects (B), and an imitative movement in fast triple time (D)—alternate with slow sections in A minor (C) and D minor (E) in the satisfying quasi-symmetrical pattern ABCBCDED.

Similarly, Purcell uses a number of strategies in the main part of the work to keep an unprecedentedly large, diverse, and wide-ranging structure under control. The key scheme is unusually varied—there are movements in eight keys—but they are laid out in a logical sequence of related keys, leading from D major/minor to the most distant point, E minor, and then back again. It is no coincidence that the E minor movement, 'Wondrous machine', is the second of the three ground basses, for they act as structural reference points, just as the three ground basses do in *Dido and Aeneas*. In the same way, the opening and closing choral sections balance one another, and are linked in various ways: they both have declamatory choral invocations to St Cecilia, and the fugal passage 'Fill ev'ry heart with love of thee' uses the same rhythmic patterns as 'Who whilst among the choir above'. Massive choral writing of this sort, inspired in part by the choruses of Draghi's 1687 ode, was something new in English music. It was the element of the Purcell style that chiefly attracted Handel, and thus it marks the true beginning of the English secular choral tradition.

Purcell's last four odes can be dealt with fairly quickly, not because they are inferior to those already discussed, but because they do not contain many innovations; once he had assimilated all the elements of the Italian style as represented by Draghi's ode, he was apparently content not to develop the genre much further. 'Celebrate this festival' Z321, a setting of Tate's poem for Queen Mary's 1693 birthday, uses the first two movements of the symphony to 'Hail, bright Cecilia', transposed from D major to C major, and is memorable mainly for the fine ground bass 'Crown the altar, deck the shrine', also known in a keyboard arrangement, ZD222. The long and flashy solo 'While for a righteous cause', for bass, trumpet, and continuo, illustrates another way in which Purcell came to terms with the da capo aria: a C major duple-time movement in declamatory style frames a triple-time passage, which plunges for a few bars into the relative minor, as in Italian arias (Ex. 5.8). In the process, he writes for the trumpet in A minor and D minor using notes available on the ordinary C instrument—a piquant effect much exploited by Austrian composers, and probably brought to England by the Moravian composer Gottfried Finger.[46]

'Great parent, hail' Z327 is one of Purcell's least-known works.

Ex. 5.8. 'While for a righteous cause' from 'Celebrate this festival' Z321/12, bars 30–41

[46] Holman, *Four and Twenty Fiddlers*, 434.

It was written for the centenary of the foundation of Trinity College, Dublin, and was performed in Christ Church Cathedral on 9 January 1694 by 'the principal Gentlemen of the Kingdom', according to John Dunton's *Some Account of my Conversation in Ireland* (1699).[47] The text is by Nahum Tate, 'who was bred up in this College', and who presumably obtained the commission for Purcell. It is not known whether the composer attended in person, though there was a performance of *Dioclesian* in London on 10 January, so it is likely that he did not.[48]

Purcell used only two recorders and strings, and avoided difficult solo writing, probably because he was not sure what performers would be available, or how good they were. By and large, too, he uses a bland Italianate idiom, with a good deal of trumpet-like passages in C major, and an inordinate number of echo effects; I wonder, not entirely in jest, whether the echoes were an instinctive response to the challenge of writing a piece about and for distant Dublin. Westrup was moved to criticize the work for 'several pages of barren pomposity',[49] but they do at least serve as a foil for the good things: the hair-raising chromatic passage in the first chorus at the words 'Who hast thro' last distress survived', and 'Awful matron, take thy seat', one of Purcell's finest declamatory solos for bass accompanied by four-part strings.

'Come, ye sons of art, away' Z323 is deservedly one of Purcell's most popular odes. It is the sixth and last of the series for the queen's birthday, and was performed on 30 April 1694; Mary died of smallpox on 28 December of that year. The work is even more tightly organized than 'Hail, bright Cecilia'. Once again there are three ground basses (G) acting as structural reference points, as well as three solo vocal and choral minuets (M). Moreover, the opening minuet is repeated after the first ground, so that the vocal movements are laid out in the following near-symmetrical pattern: MGMGMXGM. Here, X stands for the only extraneous element in the scheme, the declamatory solo 'Bid the Virtues, bid the Graces', which is thus perceived as the emotional centre of the work. It is worthy of its role, for it is a poignant piece, scored in an unusual and effective way as a duet for soprano and oboe, the latter gently echoing the former.

[47] H. Purcell, *Miscellaneous Odes and Cantatas*, ed. A. Goldsbrough, D. Arundell, A. Lewis, and T. Dart, *Works I*, 27 (London, 1957), p. xvi; see also Zimmerman, *Purcell: Life and Times*, 232–4.

[48] Van Lennep, *London Stage*, 430.

[49] Westrup, *Purcell*, 194–5.

It is strange that such a fine work should survive complete in only one source, GB-Lcm, MS 993, dating from the second half of the eighteenth century. Had an early source survived, we might be able to explain why the symphony is apparently scored just for one trumpet and one oboe when two of each are required later in the ode—are they meant to double in the symphony?—or why the final Adagio apparently leads directly into the first vocal number when Purcell usually indicated a repeat of the preceding Allegro in works of this sort. Something certainly seems to be missing: the version in *The Indian Queen* Z630/5, transposed from D major to C major, has a second oboe and a final fast movement. There is a similar problem in the symphony to Purcell's last ode, 'Who can from joy refrain?' In the ode the C major Italianate trumpet sonata appears to end with a slow C minor passage for strings alone. But a D major version of the piece was used in the incidental music for *Timon of Athens* Z632/1, and one of its sources, GB-Lcm, MS 991, indicates that it should end with the preceding fugal allegro.[50]

'Who can from joy refrain?' was written for the sixth birthday of William, Duke of Gloucester on 24 July 1695, and was probably performed at Camden House, Kensington, where his household had been established.[51] The duke, though deformed and frail, had a childish enthusiasm for all things military—he commanded a regiment of little boys, armed with wooden swords—and this is reflected in the anonymous text, and in the music. Once again the prevailing key is C major, and once again the sonorities of trumpet (only one this time, perhaps because of the limited circumstances of the first performance) and oboes predominate. Once again, too, the finest moments come in the minor-key reflective passages, particularly in the ground 'A prince of glorious race descended', constructed over an arpeggiated version of the *passacaglia*, with the first beat artfully left vacant so that the attention is focused on the stressed syllables. However, the solo 'Sound the trumpet, and beat the warlike drums', for countertenor, trumpet, and continuo, is an imaginative piece, poised midway between the declamatory and tuneful idioms, while the extended chaconne

[50] H. Purcell, *Timon of Athens*, ed. F. A. Gore Ouseley, rev. J. A. Westrup, *Works II*, 2 (London, 1974), 54.

[51] Id., *A Song for the Duke of Gloucester's Birthday (1695)*, ed. I. Spink, *Works II*, 4 (London, 1990), pp. vi–viii; see also O. Baldwin and T. Wilson, '"Who Can from Joy Refraine?": Purcell's Birthday Song for the Duke of Gloucester', *MT* 122 (1981), 596–9.

that ends the work, based on an eight-bar theme derived from the *ciaccona*, mixes ground-bass and rondeau methods of construction in a satisfying way, and brings the work—and Purcell's series of odes—to a rousing conclusion.

Although most of Purcell's odes are rarely performed today, they were evidently highly thought of in the decades after his death. The manuscript sources show that a number of them continued to be performed, while many of the solo movements were kept alive by being included in *Orpheus Britannicus*, and the symphonies were sometimes detached from their parent works and were used in the theatre or as concert pieces. Furthermore, Purcell's followers continued to write in a similar style throughout the rest of the reigns of William III and Anne; the idiom was still flourishing in 1713, when Handel produced the birthday ode 'Eternal source of light divine', his greatest tribute to Purcell. Perhaps the last word should be left to the satirist Tom Brown, who wrote of 'his unknown Friend, Mr. Henry Purcell', that 'Our whole Poetic Tribe's oblig'd to you. | For where the Author's Scanty Words have fail'd, | Your happier Graces, Purcell, have prevail'd.' He was writing in *Harmonia sacra*, ii (1693), so he was doubtless thinking of the composer's sacred songs, but his words apply with equal force to the odes.

VI
THEATRE MUSIC

ONE of Charles II's first acts on his return to London was to authorize the reopening of the theatres, closed since the beginning of the Civil War in 1642.[1] On 21 August 1660 Thomas Killigrew and Sir William Davenant, the leading survivors of the pre-war London theatre, were granted a patent allowing them to form two companies under the patronage of the king and the Duke of York. This was something new. There had been companies under court patronage in Elizabethan and Jacobean times, and Charles I granted Davenant a patent to build a theatre in 1639, a project never realized. But from 1660 the two companies enjoyed a monopoly of theatrical activity in London (it persisted until the nineteenth century, despite many vicissitudes and attempts to circumvent it), and they were supported by the king, who took an interest in their affairs, and regularly patronized them in person. His father, by contrast, had never set foot in a commercial playhouse.

The two companies started with makeshift theatres. After a spell in 1660 working together at the Cockpit in Drury Lane, an old Jacobean playhouse, Killigrew and Davenant took over two converted tennis courts, in Vere Street, and nearby in Lincoln's Inn Fields. In 1663 the King's Company opened a specially built theatre in Bridges Street, Covent Garden; it was destroyed by fire in 1672, and was replaced in 1674 by the first Theatre Royal in nearby Drury Lane. Meanwhile, the Duke's Company moved in 1671 to a specially built and lavishly equipped theatre in Dorset Garden near Charing Cross. When the companies were forced to amalgamate in 1682 they tended to reserve Dorset Garden for operatic works, which depended on elaborate spectacle, and used Drury Lane for ordinary spoken plays. Purcell's theatrical career

[1] For the history of the Restoration theatre, see L. Hotson, *The Commonwealth and Restoration Stage* (New York, 1962); A. Nicoll, *A History of English Drama*, i: *Restoration Drama, 1660–1700* (4th edn., Cambridge, 1952); and the introduction to Van Lennep, *The London Stage*.

was spent in these two buildings, which survived until the early eighteenth century.

At first, for want of anything else, the new companies revived the pre-war repertory of plays, but they soon began to produce them in a manner that owed more to the court masque—or to Italian opera, which Davenant and Killigrew had both experienced during the Interregnum. Women appeared on the stage in public for the first time, in place of the boys who had hitherto taken female parts; the theatres were equipped with changeable scenery and machines; and the companies made a determined effort to cater for the wealthy and fashionable sections of London society, as Killigrew boasted to Samuel Pepys one evening in February 1667, comparing conditions in the theatre before and after the Civil War:

That the stage is now by his pains a thousand times better and more glorious than ever heretofore. Now, wax-candles, and many of them; then, not above 3lb. of tallow. Now, all things civil, no rudeness anywhere; then, as in a bear-garden. Then, two or three fiddlers; now, nine or ten of the best. Then, nothing but rushes upon the ground and everything else mean; and now, all otherwise. Then, the Queen seldom and the King would never come; now, not the King only for state, but all civil people do think they may come as well as any.[2]

Killigrew tried to take the credit for these changes, but in fact some of them flowed from Charles II's decision to patronize the theatres in person. He was only able to boast of 'nine or ten of the best' fiddlers because he could count on the services of members of the Twenty-four Violins. Royal musicians had traditionally performed at events attended by the monarch outside the court, even those organized and paid for by the host. Thus, royal violinists probably first began to work for the two patent companies because Charles II had taken to attending their theatres, but it soon became a regular arrangement that did not depend on the king's presence. Moreover, Chapel Royal singers were occasionally allowed to take part in special productions, to judge from an order headed 'Chappell men for the theatre' relating to *The Tempest; or, The Enchanted Island*, the operatic version of Shakespeare's play produced in 1674 at Dorset Garden.[3] Arrangements of this sort seem to have lasted until the late 1670s or even later. Thereafter,

[2] Pepys, *Diary*, viii. 55–6.

[3] Holman, *Four and Twenty Fiddlers*, 331–5; Ashbee, *RECM* i. 138.

the theatres seem to have employed their own groups of musicians; we know little about them, though the establishment of a proposed United Company for, probably, the 1702–3 season allows for a 'Master to oversee the Musick', John Eccles, and 'Twenty musitians allowing near 20 sh(illings) p(er) week to each'.[4]

The presence of sizeable instrumental groups in the theatres had another consequence. They were now too large to be accommodated in small galleries above the proscenium arch or at the side of the stage, the traditional places in pre-war theatres. The solution was to put them in a pit in front of the stage, as in Continental theatres. Pepys, describing the Bridges Street theatre in May 1663, wrote that 'the Musique being below, and most of it sounding under the very stage, there is no hearing of the bases at all, nor very well of the trebles, which must sure be mended'.[5] Music rooms in galleries continued to exist throughout the Restoration period, to judge from references in plays, but operatic works must have been performed with at least some of the instrumentalists in a pit.[6] Indeed, Thomas Shadwell wrote in the text of the 1674 *Tempest* that 'The Front of the Stage is open'd, and the Band of 24 Violins, with the Harpsicals and Theorbo's which accompany the Voices, are plac'd between the Pit and Stage.'

All Restoration plays featured music. At the very least, the instrumentalists would have played two sequences of pieces before the play began (the 'first music' and the 'second music'), an overture or 'curtain tune' as the curtain rose, and 'act tunes' between the acts, as well as any dances needed. This was nothing new—there are reports of incidental music in the Elizabethan theatre—but the practice of writing suites of incidental music specially for each play started soon after the Restoration; the first surviving suite seems to be John Banister's for *The Indian Queen* by John Dryden and Sir Robert Howard, probably written for the original production in January 1664.[7]

Most Restoration plays also need vocal music of some sort, ranging from single songs and catches to extended passages of

[4] Nicoll, *A History of English Drama*, ii: *Early Eighteenth Century Drama* (3rd edn., Cambridge, 1952), 276–7; J. Milhous, 'The Date and Import of the Financial Plan for a United Theatre Company in P.R.O. LC 7/3', *Maske und Kothurn*, 21 (1975), 81–8.

[5] Pepys, *Diary*, iv. 128.

[6] C. A. Price, *Music in the Restoration Theatre* (Ann Arbor, Mich., 1979), 82–7.

[7] Price, *Music in the Restoration Theatre*, 181–2; Holman, *Four and Twenty Fiddlers*, 333–4.

operatic music. It was mostly used in situations where the audience would reasonably expect music: in drinking or seduction scenes, for serenades or lullabies, to celebrate battles or lament death, or simply for entertainment. The singers were usually cast as musicians, servants, soldiers, shepherds, priests, or supernatural beings, as they had been in pre-war plays and masques, though leading characters sometimes sang mad songs or laments, with telling effect. Singers often accompanied themselves on theorbo or guitar, avoiding the need for separate continuo players. Keyboards were not an invariable part of a theatre's equipment even after 1700, and it is likely that suites of incidental music were usually played without continuo (see below).[8]

Extended pieces of concerted music were usually reserved for two situations. Ritual scenes naturally required music, whether the protagonists were Christian or pagan priests, soothsayers, enchanters, or magicians, engaged in communal prayer, sacrificing to the gods, foretelling the future, or summoning up supernatural beings. The other situation was the presentation of self-contained masques or plays within plays. They might be presented by ordinary humans, as in Matthew Locke's Orpheus and Eurydice masque in Act IV of Settle's *The Empress of Morocco* (1673), or by supernatural characters summoned up by magic, as in the masque of Neptune and Amphitrite laid on by Prospero in Act V of the 1674 *Tempest*.[9] The two types depended on spectacular scenic effects as much as music, and the Dorset Garden theatre allowed the Duke's Company to develop them in the 1670s to the point where their parent plays became a distinct genre, called 'dramatic opera' by Dryden, 'semi-opera' by Roger North, or just 'opera' by the less precise majority. Some all-sung operas were seen in Restoration London, as we shall see, but semi-opera remained the mainstream.

The 1674 *Tempest*, with music by a consortium headed by Locke and Pelham Humfrey, is often thought of as the first semi-opera, though it only has two extended musical episodes, and its score is dwarfed by Locke's for Shadwell's *Psyche*, probably produced by the Duke's Company in February 1675.[10] *Psyche* was the first semi-opera written from scratch, and it has more than a

[8] J. Milhous and C. A. Price, 'Harpsichords in the London Theatres, 1697–1715', *EM* 18 (1990), 38–46.

[9] Locke, *Dramatic Music*, 5–16, 68–81.

[10] For a recent discussion of the date, see Holman, *Four and Twenty Fiddlers*, 346–7.

dozen musical episodes. It also requires a huge orchestra, including trumpets and drums, cornett and sackbuts, six types of woodwind instrument, strings, and continuo instruments—many of which must have been played by royal musicians. I have argued that the work was Locke's response to the arrival in the summer of 1673 of a French opera company led by Robert Cambert.[11] The Royal Academy of Music, as it was called, associated itself with the King's Company, and produced the opera *Ariane, ou le mariage de Bacchus* at Drury Lane in March 1674. It did not last long, but the French wind-players who came over with Cambert, apparently bringing with them the first Baroque oboes and recorders to be heard in England, stayed on at Drury Lane for a while, giving the King's Company a temporary advantage in musical matters over the Duke's Company—which had hitherto been the more musical of the two.

The Duke's Company put on one more semi-opera—Charles Davenant's *Circe*, set by Banister and produced in May 1677—before circumstances brought the genre to a halt. Locke and Banister, the main theatre composers, died in August 1677 and October 1679, and the King's Company, weakened by internal disputes and by the contemporary political crisis, was forced to amalgamate with the Duke's Company; a United Company was formed in November 1682. For one reason or another, the 1680s were depressed years for music in the commercial theatre, the more so after *Albion and Albanius* failed in 1685. This unlucky work, Louis Grabu's all-sung French-style opera on an allegorical text by Dryden praising Charles II and his brother, was in rehearsal when Charles died in February, and a revised version was abandoned after a few performances in June when the news of Monmouth's invasion reached London. The production of semi-operas was only resumed with Purcell's *Dioclesian* Z627, produced by the United Company in June 1690.

This, then, was the situation when Purcell wrote his first theatre music, for Nathaniel Lee's *Theodosius* Z606, produced at Dorset Garden in the spring or summer of 1680.[12] Some of it was published anonymously in the play text, but his authorship is confirmed by John Downes, the prompter of the Duke's Company, who wrote that it contained 'several Entertainments of

[11] Holman, *Four and Twenty Fiddlers*, 343–53.

[12] For the date, see Price, *Henry Purcell*, 30.

Singing; Compos'd by the Famous Master Mr. Henry Purcell, (being the first he e'er Compos'd for the Stage)'.[13]

Purcell's main contribution was a religious ceremony in Act I, scored for solo and tutti voices with two recorders and continuo. The early Christian emperor Theodosius has decided to retire to a monastery, requiring his sisters to accompany him. A trio of priests ask the women in a series of anthem-like passages if they are ready for the monastic life, to which Marina and Flavilla reply in what initially sounds like two verses of a strophic song; in fact, they soon diverge, and Marina's has the more tortured melodic line, perhaps because she is the one with the doubts. They urge the priests to 'Haste then, O haste then, and take us in', and then the recorders start what sounds like a French overture, but a priest joins in for the repeat of the first section, and Purcell adds the whole choir for the second, triple-time section. This striking device parallels the choral repeat of the second section of the symphony in 'Welcome, vicegerent of the mighty king' Z340, written that same summer; it was perhaps used to suggest that the monastic life is an overture to heavenly joys. Purcell avoids elaborate counterpoint and complex writing in the scene, presumably to evoke the simplicity of early Christian ceremonies, but keeps up the interest with a stream of Locke-like melodic and harmonic twists and turns.

Theodosius was a success. Downes wrote that the acting and the music 'made it a living and Gainful Play to the Company: The Court; especially the Ladies, by their daily charming presence, gave it great Encouragement'. It is odd, therefore, that Purcell's next two theatre works were for the troubled King's Company. He wrote the song 'Retir'd from any mortal's sight' Z581 for Nahum Tate's *The Sicilian Usurper*, an adaptation of Shakespeare's *Richard II* (?December 1680), and 'Blow, Boreas, blow' Z589 for D'Urfey's comedy *Sir Barnaby Whigg* (?October 1681).

After *Sir Barnaby Whigg* Purcell composed little for the London theatres for nearly a decade. Some works have been wrongly allocated to this period. The suite for *The Double Marriage* Z593 (dated '?1682–5' by Zimmerman) is now known to be by Grabu,[14] while *Circe* Z575 ('1685'), may actually date from

[13] J. Downes, *Roscius Anglicanus* (London, 1708), ed. J. Milhous and R. D. Hume (London, 1987), 80.

[14] Price, *Henry Purcell*, 14.

around 1690 (see below). Similarly, 'Beneath the poplar's shadow' Z590 for Lee's *Sophonisba* ('?1685') seems to have been written for a 1690s revival.[15] This only leaves music for four plays: the song 'How great are the blessings of government made' Z494 for Tate's *Cuckholds-Haven* (July 1685); the catch 'My wife has a tongue' Z594 for Edward Ravenscroft's *The English Lawyer* (?1684–5); eight songs for D'Urfey's *A Fool's Preferment* Z571 (April 1688); and the first setting of 'Thy genius, lo!' Z604, for Lee's *The Massacre of Paris* (?November 1689).

We do not know whether this was because Purcell deserted the theatre, or the theatre deserted Purcell. Curtis Price argues it was the latter. He points out that the United Company concentrated on revivals rather than new works, which would have required little or no new music, and suggests 'his leaving Dorset Garden after *Theodosius* shut him out from working for the United Company, which turned instead to [Simon] Pack for most of the music in its new productions'.[16] On the other hand, it is hard to believe that the managers of the United Company would have preferred Pack to Purcell, however many bad career moves Purcell had made. Perhaps he was just too busy. In the early 1680s he was turning out vocal works of all sorts as part of his court duties, and was probably involved in performances at Whitehall on more or less a daily basis. He wrote less for the court during the reign of James II, but he was still essentially a court musician until the Glorious Revolution. In the spring of 1690 the activities of the royal music were cut back, and he and his colleagues were free to pursue other interests. At that moment he began a hectic new career with the United Company.

Relevant to this discussion, of course, is the date of *Dido and Aeneas* Z626. Until recently it was accepted that the work was written for the boarding-school for girls run in Chelsea by the dancing-master Josias Priest, and that it had been performed there in the spring of 1689; the evidence is a unique printed copy of Tate's text, as well as D'Urfey's spoken epilogue, published in his *New Poems* (1690).[17] The first challenge to this orthodoxy came unexpectedly in 1989, when a printed text of Blow's *Venus and Adonis* relating to a performance at Priest's school in 1684 came to

[15] Price, *Henry Purcell*, 58–9. [16] Ibid. 13.

[17] There is a facsimile of the text in H. Purcell, *Dido and Aeneas*, ed. A. M. Laurie, *Works II*, 3 (London, 1979), pp. xiii–xx; there is a critical edition in Purcell, *Dido and Aeneas*, ed. C. A. Price (New York and London, 1986), 63–79.

light.[18] Since *Venus and Adonis* was first performed at court in about 1682, it soon occurred to scholars that the same thing might have happened to *Dido*—that is, it might have been written several years earlier for an unrecorded court production. The idea has obvious attractions, not least because it explains why a work supposedly written for girls needs so many male singers: Aeneas, the Sailor, the Sorceress (which, it has been argued, was conceived as a travesty role for a bass—or rather, a baritone[19]), and three of the four chorus parts—the alto can only have been sung as it stands by high tenors: it is generally lower than normal, and in 'Haste, haste to town' it goes down several times to *d*, and is rewritten in some modern editions.

So far, the problem has been tackled from two angles. Andrew Pinnock and Bruce Wood have argued largely on grounds of musical style that *Dido* was written in about 1684.[20] Their case has its merits, though it is notoriously difficult to argue from stylistic evidence in this way, particularly when it depends on changes of style over such a short period. By contrast, Andrew Walkling has argued that *Dido* is an allegory of events in the reign of James II; or rather, that it is a warning, delivered in 1687 or early 1688, to the Catholic king (Aeneas) not to be duped by the bad counsel of the witches (English papists) into abandoning Dido (a personification of England) for Rome.[21] This certainly makes more sense than the usual interpretation of *Dido* as an allegory of the Glorious Revolution, which, as Curtis Price has pointed out, founders on the identification of William III as Aeneas: 'the story of a prince who seduces and abandons a neurotic queen would seem a tactless way to honour the new monarchs'.[22] Many of the points made by Pinnock and Wood would still be valid for a work written just before the Italianate stylistic revolution initiated in part by Draghi's 1687 St Cecilia Ode. I suspect, however, that the last word has not been said on the matter.

[18] Luckett, 'A New Source for *Venus and Adonis*'.

[19] I. Cholij and C. A. Price, 'Purcell's Bass Sorceress', *MT* 127 (1986), 615–18.

[20] Pinnock and Wood, '"Unscarr'd by Turning Times"'; see also M. Adams, 'More on Dating *Dido*', *EM* 21 (1993), 510; C. A. Price, '*Dido and Aeneas*: Questions of Style and Evidence', *EM* 22 (1994), 115–25; Pinnock and Wood, '"Singin' in the Rain": Yet More on Dating *Dido*', ibid. 365–7.

[21] A. R. Walkling, 'Politics and the Restoration Masque: The Case of *Dido and Aeneas*', in G. Maclean (ed.), *Literature, Culture, and Society in the Stuart Restoration* (Cambridge, forthcoming).

[22] Price, *Henry Purcell*, 229; see J. Buttrey, 'Dating Purcell's *Dido and Aeneas*', *PRMA* 94 (1967–8), 51–62.

The other major problem concerning *Dido and Aeneas* is the form in which it survives. The earliest source, GB-Ob, Tenbury MS 1266 (*c*.1775), lacks some sections present in the 1689 text: the prologue, a chorus at the end of Act II, and some of the dances.[23] Some scholars have argued that the work is essentially complete, but most have accepted that there really is music missing, and there have been a number of attempts to reconstruct the lost chorus and dances by adapting other music by Purcell, though the Prologue is probably irretrievably lost.[24] The explanation, first proposed by Eric Walter White and refined and developed by Curtis Price, is that the Tenbury *Dido* is truncated in part because it was inserted by Charles Gildon into his adaptation of *Measure for Measure*, put on at Lincoln's Inn Fields in 1700.[25] Gildon and his musical collaborator (?John Eccles) chopped up *Dido* into four separate entertainments, placing the Prologue last, reversing the order of the Palace and Cave scenes (despite the damage done to the dramatic logic), and cutting out some of the dances. In 1704 *Dido* was performed at Lincoln's Inn Fields in the company of Ravenscroft's *The Anatomist* and Etherege's *The Man of Mode*. At this point, it is argued, the Palace and Cave scenes were restored to their original order, though the Prologue was omitted, and the other cuts were not restored, producing essentially the version that has come down to us.

Dido and Aeneas was probably never performed in public in the composer's lifetime, and does not seem to have made much of an impression, even on Charles Gildon, who must have known it well. He did not mention it when he praised Purcell in his *Life of Mr. Thomas Betterton* (London, 1710), p. 167, though he compared several pieces from the semi-operas favourably with the music of the Italian operas put on in London in the first decade of the eighteenth century: 'Let any Master compare Twice ten hundred Deities, the Music in the Frost Scene, several Parts of the Indian Queen, and twenty more Pieces of Henry Purcel, with all the Arrieto's, Dacapo's, Recitativo's of Camilla, Pyrrhus,

[23] For the date of the Tenbury manuscript, see E. T. Harris, *Henry Purcell's Dido and Aeneas* (Oxford, 1987), 45.

[24] See Price, *Henry Purcell*, 241–2 for details; there is a necessarily speculative reconstruction of the Prologue by Margaret Laurie in Purcell, *Dido and Aeneas*, ed. T. Dart and Laurie (London, 1961).

[25] E. W. White, 'Early Theatrical Performances of Purcell's Operas', *Theatre Notebook*, 13 (1959), 2–24; id., 'New Light on *Dido and Aeneas*', in Holst (ed.), *Purcell: Essays on his Music*, 14–34; Price, *Henry Purcell*, 234–8.

Clotilda, &c. and then judge which excels.' Similarly, Roger North, who knew Purcell and played with him on several occasions (see Ch. 3), discusses several pieces from the semi-operas with enthusiasm, including the Frost Scene from *King Arthur*, but does not mention *Dido*.[26]

We find this hard to understand, but that is because our culture regards all-sung opera as the principal type of music theatre. Works that mix speech and music tend to be thought less important, at least by opera critics. But the spoken theatre was the mainstream in seventeenth-century England, and the men who ran the commercial playhouses saw no reason to put on all-sung works, particularly when it involved ordinary mortals singing when there was no rational reason for them to do so. 'Irrational' behaviour of this sort was considered acceptable for personifications, allegorical characters, and gods and goddesses; a work concerned with them, Dryden pointed out in a famous phrase in his preface to *Albion and Albanius*, 'admits of that sort of marvellous and surprizing conduct, which is rejected in other Plays'.[27]

It is no accident that the few all-sung works produced in seventeenth-century England belong to an offshoot of the masque, the home of gods and goddesses, and 'marvellous and surprizing conduct'. Court masques, essentially vehicles for the display of courtly dancing, always mixed speech with music, but a private version of the form developed in country houses in the reign of James I, and in these intimate surroundings, freed from the demands of flattery and ostentatious display, playwrights and composers were able to experiment: Ben Jonson's *Lovers Made Men*, produced at Essex House in 1617, was apparently all-sung, set by Nicholas Lanier *stylo recitativo* 'after the Italian manner'.[28]

There were no more court masques after 1642, and so the private masque became the mainstream. At the same time, unemployed playwrights and actors were looking for ways to circumvent Parliament's ban on the public performance of stage plays. A solution, tried by Davenant with *The Siege of Rhodes* (prepared in 1656, but probably not performed until 1659), was to

[26] Wilson, *Roger North*, 217–18, 220–1, 307, 353.

[27] J. Dryden, *Works*, 15: *Plays: Albion and Albanius, Don Sebastian, Amphitryon*, ed. E. Miner, G. R. Guffey, and F. B. Zimmerman (Berkeley, Los Angeles, and London, 1976), 3.

[28] Spink, *English Song*, 46.

disguise a spoken play by having it sung in recitative, or at least to add enough music to fool the authorities—which was probably why Locke expanded the role of the music in Shirley's masque *Cupid and Death*, setting a number of the original speeches as recitative.[29] This was done in 1659 for a production in a London riding-school, which might have been construed as a public performance.

After the Restoration there was no need for all-sung works in the public theatres—Davenant produced *The Siege of Rhodes* as a spoken play[30]—but the private masque continued, particularly in schools, and around 1682 John Blow produced *Venus and Adonis* for a court performance. Blow's *Venus and Adonis* is entitled 'A Masque for the Entertainment of the King' in the primary source, and it belongs to the private masque tradition in that it was performed partly by amateurs—Venus and Cupid were played by Charles II's mistress Moll Davies and their young daughter Lady Mary Tudor—and that there was much dancing: there are eight dances, half of them grouped at the end of Act II.[31]

It is not clear why Blow and his anonymous librettist decided to create an all-sung work rather than a traditional masque, though it probably had a lot to do with Charles II's interest in exotic entertainment. The king made several attempts to establish an Italian opera company in England in the 1660s, and in 1676 he sent Nicholas Staggins to France and Italy for two years, apparently to study operatic practice there.[32] The fruit of Staggins's trip was an operatic venture in collaboration with Blow. He probably would have overseen the production of *Venus and Adonis* as Master of the Music, and on 4 April 1683 the king responded to a lost petition from the two musicians 'for the erecting an Academy or Opera of Musick, & performing or causing to be performed therein their Musicall compositions'.[33] The outcome is unknown, but Pinnock and Wood have argued that it provided Purcell with the stimulus to write *Dido and Aeneas*.

[29] M. Locke and C. Gibbons, *Cupid and Death*, ed. E. J. Dent (MB 2; 2nd edn., London, 1965).

[30] J. Protheroe, 'Not so much an Opera . . . a Restoration Problem Examined', *MT* 106 (1965), 666–8.

[31] J. Blow, *A Masque for the Entertainment of the King: Venus and Adonis*, ed. C. Bartlett (Wyton, 1984).

[32] Mabbett, 'Italian Musicians in Restoration England'; Holman, *Four and Twenty Fiddlers*, 302–4.

[33] GB-Lpro, SP44/55, 248; there is a facsimile of the document in Wood and Pinnock, '"Unscarr'd by Turning Times"', 387.

Dido was certainly modelled on *Venus and Adonis*. Tate and Purcell followed Anon. and Blow in laying it out with a prologue and three acts; in casting the protagonists as a soprano and a baritone or bass; in giving them extended passages of expressive declamatory dialogue; in giving the chorus a prominent role, and requiring them to dance (the 1689 text mentions seventeen dances); in mixing comedy with tragedy; in using the motifs of the hunt and the wild boar; in using a key scheme that progresses to the death key of G minor by way of F and D key centres; and by ending the work with a situation of great pathos, set to heart-rending music. Most important, perhaps, both operas have the same distinctive tone, created by the brevity of the movements, and the speed with which the action hurtles towards the inevitable tragic conclusion. In *Dido and Aeneas* this is intensified by Purcell's use of chains of tiny contrasted sections, linked by proportional time signatures; the Cave Scene, for instance, lasts only a few minutes, and consists of nine sections intended to be performed at a constant pulse, so that they sound like a single movement.[34]

Much has been made of the supposed French or Italian influence on *Venus and Adonis* and *Dido and Aeneas*. In general terms, it is probably true that their declamatory writing was modelled to some extent on the recitative of mid-century Venetian operas, just as their prologues and their dance-based airs and choruses may owe something to French models. Cavalli's *Erismena* was evidently performed or prepared for performance in Restoration England, for a contemporary manuscript score survives with English text, and there were some performances of French operatic works in London, including, supposedly, Lully's *Cadmus et Hermione* in 1686.[35]

But, as so often in English music, foreign models seem to have influenced the planning of these operas rather than their musical style. Blow and Purcell did not need to imitate Cavalli's type of recitative, for they had plenty of models in Locke and earlier English song composers, though Venetian opera might have helped them and their librettists to develop an Italianate relationship between the declamatory passages and the concerted

[34] Proportional time signatures are suggested in Purcell, *Dido and Aeneas*, ed. Laurie; see also Purcell, *Dido and Aeneas*, ed. E. Harris, full score (Oxford, 1987), pp. vii–viii; and my review in *ML* 71 (1990), 617–20.

[35] T. Walker, 'Cavalli', *Grove 6*, iv. 32; Van Lennep, *London Stage*, 347.

numbers, based on the contrast between action and reflection, conflict and the portrayal of character. Similarly, they did not need French operas to know how to use dance forms as the basis of extended sequences of solos and choruses, as in the minuets 'Fear no danger to ensue' and 'Thanks to these lonesome vales', as structures of this sort had long been a feature of court odes.

The main formal innovation of *Dido and Aeneas*, not found in *Venus and Adonis*, was the use of three vocal ground basses, placed at strategic points in the drama. They help to articulate and define the most important key centres, C minor, D minor, and G minor, and they serve as structural pillars at the beginning, middle, and end of the work, as in some odes. Each one has a different dramatic function. The first, 'Ah! Belinda', a minor-key variant of the *ciaccona*, reveals the depths of Dido's guilty passion for Aeneas, and its tragic yet sensuous music immediately establishes her as a well-rounded character. 'Oft she visits this loved mountain' is an example of the duple-time ground in flowing quavers often found in the odes of the early 1680s. This is appropriate, for it is sung by one of Dido's courtiers in an ode-like situation, entertaining Dido and Aeneas, and it creates the necessary moment of repose before the storm and the appearance of the false Mercury. Dido's Lament, a chromatic *passacaglia*, the most potent emblem of love and death in Italian music, raises the tone of the work on to a much grander, richer plane than Blow aspired to in *Venus and Adonis*.

Incidentally, Tate's text allows for two ground-bass dances accompanied by guitar, the 'Dance Gittars Chacony' in the Palace Scene, and the 'Gittar Ground a Dance' in the Grove Scene. There is no music for them (they were probably improvised by the guitarist) but they need to be in C major and D minor, and therefore may have been sets of strummed variations on the *ciaccona* and the *passacaglia*, which are often found in these keys. Perhaps Purcell planned *Dido* with a guitarist in mind, such as Nicola Matteis, whom Roger North described as a 'consumate master' on the guitar, who 'had the force upon it to stand in consort against an harpsicord'.[36] It is striking how many numbers are ideally designed for a strummed guitar continuo: one thinks of the fugue of the overture, 'Banish sorrow, banish care', 'When monarchs unite', 'Fear no danger to ensue', 'Pursue thy conquest,

[36] Wilson, *Roger North*, 357.

Love', 'To the hills and the vales', the Triumphing Dance, the 'Ho, ho, ho' choruses, 'Come away, fellow sailors', 'Destruction's our delight', and so on.

Purcell's career in the commercial theatre began in earnest in 1690. He wrote more than forty theatre works over the next five years, ranging in scope from single songs to complete semi-operas. They come so thick and fast at this period, and there is so much doubt about their precise dating, that it will be convenient to abandon the chronological narrative at this point. The three semi-operas he wrote in 1690–2 are best considered together, as are the various types of smaller theatre works.

The Prophetess; or, The History of Dioclesian, first performed at Dorset Garden in early June 1690, is an operatic adaptation of a Beaumont and Fletcher play, probably made by Thomas Betterton, the leading figure in the United Company. Set in the late Roman empire, it concerns a prophecy made by Delphia (the prophetess) that the soldier Diocles will become emperor by killing a boar. For a while he concentrates on real beasts, but in Act II it becomes evident that his proper target is Aper (= boar), the captain of the guard, who has been accused of the murder of the previous emperor by the present incumbent, Charinus. His reward is half the empire, and the princess Aurelia—which involves ditching his betrothed Drusilla, Delphia's niece. This enrages Delphia, who engineers his defeat at the hands of the Persians. The chastened Diocles repents, and is allowed a victory before he abdicates in favour of his nephew Maximian (who promptly plots to kill him); at the end of the play he retires to Lombardy with Drusilla, to be welcomed by an extended masque of Cupid and Bacchus.

There have been attempts to read *The Prophetess* as a political allegory, despite the fact that it was an old play to which Betterton made only minor changes. John Buttrey equates William III with Diocles and James II with Aper, while Curtis Price identifies James with Diocles, Monmouth with Maximian, Charles II with Charinus, Cromwell with Aper, and Charles I with his victim.[37] So far, so good, but James did not abdicate in favour of Monmouth, and Monmouth's rebellion did not end in his pardon. Also, it is hard to imagine that a work featuring a James II figure as the hero would have been successful in 1690, even if he was

[37] J. Buttrey, 'The Evolution of English Opera between 1656 and 1695: A Reinvestigation' (Ph.D. thesis, Cambridge, 1967), esp. 243; Price, *Henry Purcell*, 270–2.

presented, in Price's words, as 'a fop and a fool, whose humility is forced on him by stratagem'. Price is on surer ground when he sees the work's 'gallery of flawed characters' as 'a welcome antidote to Dryden's nauseatingly patriotic *Albion and Albanius*'.

Dioclesian (as the score is usually known) is a landmark in the history of English theatre music. It was the first semi-opera since the 1670s, and the first theatre work by an Englishman to be published in full score (Locke's *Psyche* appeared incomplete in a compressed format). Grabu imitated Ballard's handsome full scores of Lully's operas when he published *Albion and Albanius* in 1687, which Purcell imitated in turn. A large number of copies of *Dioclesian* survive, but it does not seem to have been a commercial success. John Walsh wrote in a preface to the score of Daniel Purcell's *Judgement of Paris* (1702) that 'the ill Success which Publishers of Musick have mett with in other approved pieces, might have (as hitherto it has been) a sufficient Discouragement from such an Undertakeing', adding that 'the Celebrated Dioclesian of Mr. Henry Purcell is an Instance of that Nature, which found so small Encouragement in Print, as serv'd to stifle many other Intire Opera's, no less Excellent, after the Performance, not Dareing to presume on there own meritt how just soever, nor hope for a better Reception then the former'.[38]

As a result, *Dioclesian* was Purcell's only semi-opera to be printed at the time, though it is by far the least known today. Price's explanation is that the tone of the lively and satirical play does not mesh with the music, which is mainly military and ceremonial in character, and that 'many of the numbers lack the boldness, the "forgèd feature" of the other semi-operas'.[39] On a practical level, it depends to a large extent on extravagant and expensive stagecraft, so stage revivals have been few and far between.

Nevertheless, *Dioclesian* contains a good deal of beautiful music, particularly in the instrumental pieces that accompany the dazzling effects. The score contains two main operatic episodes. The first comes in Act II, and starts with music conjured up by Delphia to celebrate Diocles's victory over Aper; his coronation ceremony follows. The text, triumphalist in tone, recalls court odes, and this prompted Purcell to produce an ode-like sequence

[38] On an extra leaf in Robert Spencer's copy; I am grateful to him for drawing it to my attention.

[39] Price, *Henry Purcell*, 288.

of music, starting with 'Great Diocles the boar has kill'd', a superb declamatory bass solo, richly and elaborately accompanied by four-part strings. The rest is not so good, but there is an expressive G minor soprano air with two recorders, 'Charon the peaceful shade invites', which subtly alludes to the *passacaglia* ground without using it directly. The Italianate symphony that follows after a short recitative is of interest in that it is scored for two trumpets with two violins and continuo. String-writing without violas in Purcell's major concerted works is usually a sign of the use of solo violins, and this may have been because the trumpets were being played at a distance—perhaps in a stage machine. Confirmation that at least some of the instruments were airborne is provided by the stage direction 'Then a Symphony of Flutes in the air' before the alto air 'Since the toils and the hazards of war's at an end', which has an extended prelude for two recorders and continuo.

The scene takes an extraordinary turn at the end. Diocles is offered Aurelia's hand, which he accepts, whereupon Delphia summons up 'a dreadful monster', which comes 'from the further end of the scenes and moves slowly forward', accompanied by Purcell's creepy Locke-like 'Soft music before the dance'. Then 'They who made the monster separate in an instant, and fall into a figure, ready to begin a Dance of Furies.' The music is full of the rushing scales used in French dances for demons and furies, such as the one in Grabu's incidental music for Rochester's *Valentinian* (1684), which Purcell may have heard.[40] This evocative music is matched by a *coup de théâtre* in Act IV, where the same prelude precedes another demonstration of Delphia's magic. She conjures up an image of Aurelia lying dead in a tomb, only to change it into butterflies as Diocles approaches; they dance to the sound of ethereal yet searchingly harmonized music. While we are on the subject of the instrumental music in *Dioclesian*, mention must be made of the chaconne 'Two in one upon a Ground' played 'in the third Act'. It is an exact canon for two recorders over a variant of the *passacaglia*, and is one of the most perfect examples of how contrapuntal ingenuity can be powerfully expressive.

The extended masque that ends *Dioclesian* was one of Purcell's most popular works, which might surprise someone who had no

[40] Holman, *Four and Twenty Fiddlers*, 378–81.

idea of the spectacular visual treats in store for audiences of the time. There was a vast machine filling the theatre 'from the frontispiece of the stage to the further end of the house' representing the palaces of Flora, Pomona, Bacchus, and the Sun God; it was in four tiers, with singers and dancers on each level, like some extravagant 1930s musical.[41] Much of the music is not especially memorable, though the interest is kept up by constant changes of rhythmic and harmonic direction, and by the varied scoring: strings for Cupid, trumpets and strings for the 'entry of Heroes', oboes for Bacchus; the full orchestra is reserved for the chaconne 'Triumph victorious Love', with its spectacular antiphonal writing for trumpets, double reed quartet, and strings—probably the first time such effects had been heard in a London theatre. There is one wonderful number, the soprano duet 'Oh, the sweet delights of love'. Purcell tended to excel in rondeau form, in gavottes, and in writing for pairs of soprano voices, and this seductive piece combines all three features: it has a theme full of delightful hocket-like cooings on the words 'Oh' and 'who', and surprisingly wide-ranging and varied episodes, which powerfully suggest the all-embracing nature of love (Ex. 6.1).

King Arthur Z628, Purcell's second semi-opera, was produced at Dorset Garden just under a year later, in May 1691. Like *Dioclesian*, it has a connection with *Albion and Albanius*. Dryden began to write the text in 1684, possibly for the twenty-fifth anniversary of the Restoration in May 1685.[42] It was to consist of a three-act semi-opera with an all-sung allegorical prologue. For some reason, Dryden and Grabu shelved it, and developed the prologue into a complete work, *Albion and Albanius*, performed in 1685. We do not have the text of the 1684 semi-opera, but several scholars have argued that it was originally concerned with the events of the Exclusion Crisis, with Arthur and the Britons as Charles II and the Tories, and Oswald and the Saxons as Shaftesbury and the Whigs.[43] By the time Purcell set it, it is argued, Dryden had managed to adapt it to the new political situation, with Arthur as William III, and Oswald as the deposed James II.

[41] Price, *Henry Purcell*, 282–8; J. Muller, *Words and Music in Henry Purcell's First Semi-Opera, Dioclesian* (Lewiston, NY, Queenston, Ont., and Lampeter, Dyfed, 1990), 301–16.

[42] Suggested by A. M. Laurie, 'Music for the Stage II: From 1650', in Spink (ed.), *Seventeenth Century*, 321.

[43] Buttrey, 'The Evolution of English Opera', 248–60; Price, *Henry Purcell*, 290–5.

Ex. 6.1. 'Oh, the sweet delights of love' from *Dioclesian* Z627/30, bars 1–18

Ex. 6.1. *cont.*

The plot owes little to history or even medieval romance. It begins as Arthur goes into battle to recapture Kent, the last part of the kingdom held by the Saxon king Oswald. The Saxons sacrifice to the gods, but are defeated, and resort to sorcery. In Act II the evil spirit Grimbald tries to lead the Britons into a bog, but the repentant spirit Philidel rescues them. The Saxons capture the blind Emmeline, Arthur's betrothed, imprisoning her in a magic castle. Merlin is unable to release her, but restores her sight; meanwhile, her gaoler Osmond (Merlin's evil counterpart) makes advances, conjuring up the Frost Scene, a masque of Cupid and a Cold Genius, to awaken her desire. Arthur attempts another rescue in Act IV, and Osmond tries to ensnare him with visions of beauty in an enchanted grove, but he breaks the spell and releases Emmeline. In Act V Arthur defeats Oswald in single combat, and expels the Saxons, but Merlin prophesies that the two peoples will become a united, prosperous nation, and offers them a vision of the future in the final masque.

There is much more music in *King Arthur* than in *Dioclesian*, and it is more successfully integrated into the plot; unusually, Grimbald and Philidel sing as well as speak, linking play and music. Again, most of it is in extended operatic episodes: the Sacrifice and Battle Scene in Act I; the scene in Act II where Grimbald tries to lead the Britons astray; the Frost Scene in Act III; and the Act V masque. Unfortunately, the sources are poor. Purcell did not publish the score, and there is no authoritative complete manuscript. There are some missing numbers, and we can only guess at the function of some of those that do survive. It is possible, for instance, that the D minor overture and the D major overture with trumpets are alternatives; the latter was borrowed from the ode 'Arise, my Muse' Z320/1, perhaps because the former was not ready in time for the first performance. Conversely, the ground-bass setting of 'Saint George, the patron of our isle', for soprano, two trumpets, and continuo, is almost certainly spurious, and was probably added for one of the revivals late in the 1690s. With one exception, we do not know where the act tunes should be placed, and there is even some doubt about the order of events in the Act V masque.

King Arthur certainly has editorial problems, but it has always been popular. It was Purcell's most successful stage work, was regularly revived in the next century, and held the stage in a mutilated form until the 1840s. It is easy to see why. Its music may not be consistently finer than the others, but it is wonderfully varied, and for this Dryden should take some of the credit. He understood, as Betterton failed to do, that the composer needed a series of varied situations to produce the necessary contrasts of mood and idiom in the music without distorting its relationship with the parent play.

This may seem an obvious point, but it explains why there is so much grand but rather bland music in *Dioclesian*. By the end of Act III of *King Arthur*, by contrast, one has experienced the solemn yet urgent music of the Sacrifice Scene, with its wonderful plunge into F minor at the words 'Die and reap the fruit of glory'; the heroics of 'Come if you dare', with its stirring battle music; the ethereal spirit music of 'Hither this way', with its remarkable double-choir writing (somewhat garbled in the sources); the serene minuet 'How blest are shepherds'; the rustic jollity of 'Shepherd, leave decoying'; and the brilliance and fantasy of the Frost Scene. No one would pretend that it is a perfect play. Curtis Price diagnoses its 'perceived uneveness' as 'the result of Dryden's having transformed what was originally a heartfelt parable of royal reconciliation into a backhanded compliment to a king for whom he did not much care'.[44] But there are worse things in some familiar opera librettos, and there is no reason why it should not appeal to an audience that can accept the heroic mode of Handel's operas.

The Frost Scene has always attracted praise. Roger North thought Charlotte Butler's singing of Cupid's 'recitativo of calling towards the place where Genius was to rise' 'beyond any thing I ever heard upon the stage', particularly because of 'the liberty she had of concealing her face, which she could not endure should be so contorted as is necessary to sound well, before her gallants, or at least her envious sex'.[45] Thomas Gray, who saw it in 1736, thought the Frost Scene 'excessive fine', and the Cold Genius's solo 'the finest song in the Play'.[46] We now know that Purcell modelled the famous shivering music on Lully's for the 'Peuples des climats glacés' in Act IV of *Isis* (1677), but his daring chromatic harmonies transform the Cold Genius from the picturesque figure of Lully (or Dryden, for that matter) into a genuinely awe-inspiring character—the more so because Cupid's responses are set to such frothy and brilliant music.

Furthermore, the scene is so satisfying because its diverse music is so tightly organized. The Prelude, the opening gesture of a French overture transformed by Italianate semiquavers, has its counterpart in the irresistible ritornello near the end. Cupid summons the Cold Genius and the Cold People in two recitatives, and their shivering responses, in C minor rather than the prevailing C major, are associated with matching instrumental music, a prelude and a dance respectively. The minuet-like prelude in which the Cold People appear is related in rhythm to 'Thou doting fool'; later it is used for his solo ''Tis I that have warmed thee', and (ingeniously combined with the ritornello) for their responses. In fact, the serene duet 'Sound a parley', in which the protagonists are reconciled, is the main section not connected in some way to another, and is thus felt to be the still centre of the scene.

The Act V masque is diffuse and incoherent by comparison, though this may partly be the fault of the sources. It starts magnificently, with Aeolus 'in a Cloud above' ordering the winds to retire in Locke-like storm music; then the key changes from C major to C minor, and two recorders replace the strings for the

[44] Price, *Henry Purcell*, 318. [45] Wilson, *Roger North*, 217–18.

[46] T. Gray, *Correspondence*, i: *1734–1755*, ed. P. Toynbee and L. Whibley (Oxford, 1935), 37.

couplet 'Serene and calm and void of fear | The Queen of Islands must appear'. Britannia rises, seated on an island, to another superb piece, a grave yet florid symphony for three trebles and continuo; it is said in one source to be for three violins, but the range and character of the parts suggests a trumpet and two violins, or possibly trumpet, violin, and oboe. Thereafter, the masque degenerates into a series of turns: the strangely ecclesiastical ATB trio 'For folded flocks' (an anthem to sheep?) is followed by the comic 'Your hay it is mowed' (sung by Comus and three peasants), which in turn is followed incongrously by 'Fairest isle' (sung by Venus). It would probably be effective enough in an imaginative and spectacular stage production, but it must be said that it is a strangely bathetic ending to the work, with little of the coherence of the Cupid and Bacchus masques in *Dioclesian* and *Timon of Athens*.

The Fairy Queen Z629, Purcell's third semi-opera, was first performed on 2 May 1692. It was much more expensive than *King Arthur*: 'the clothes, scenes, and musick' cost £3,000 according to Narcissus Luttrell, and Downes wrote that 'the Expences in setting it out being so great, the Company got very little by it'.[47] It is known to have been revived only once, in February 1693, before the full score was lost, putting an end to more complete performances. The managers of the Theatre Royal advertised for its return twice in 1701, apparently without result.[48] Nothing more was heard of it until it came to light in GB–Lam, just in time to be used for the 1903 edition, *Works I*, 12. *The Fairy Queen* is an anonymous adaptation of *A Midsummer Night's Dream*, which means that it has attracted more attention in this century than Purcell's other semi-operas, though most productions have tried to combine the music (which does not set a line of Shakespeare) with portions of the original play. More recently, Roger Savage and others have argued forcefully that the 1692 version of the text is the appropriate vehicle for the music, and works well in its own terms.[49]

The survival of the partly autograph Royal Academy score should have made the task of establishing a musical text for *The Fairy Queen* simple enough, but in fact the work has its own share of problems. Most of them concern the relationship of the score to the two printed texts, the quartos of 1692 and 1693. It is

[47] Luttrell, *Brief Relation*, ii. 435; Downes, *Roscius Anglicanus*, 89.
[48] Tilmouth, 'Calendar of References', 40.
[49] R. Savage, 'The Shakespeare–Purcell *Fairy Queen*', *EM* 1 (1973), 200–21.

usually said that it was prepared in connection with the 1693 revival, but Bruce Wood and Andrew Pinnock have recently argued that it dates from the original 1692 production, which means that the Act I scene with the Drunken Poet was part of the original conception even though it does not appear in the 1692 text; they suggest that Thomas D'Urfey, the stuttering poet who partially inspired it, contributed to it directly, perhaps writing the text or even acting the role himself.[50] Titania's fairies lead him in blindfolded with two silent companions, and torment him until he admits to being a poor scurvy poet. Purcell portrays the irrationality of drunkenness in the same way that he portrays madness—with a number of sudden changes of rhythmic direction—though the music has a delightful lightness that tells us that the malady is as temporary as the tormentors are insubstantial. At the end the poets are laid to rest with a delicious series of false relations over a dominant pedal (Ex. 6.2).

Ex. 6.2. 'Drive 'em hence' from *The Fairy Queen* Z629/51, bars 180–3

[50] A. Pinnock and B. Wood, '*The Fairy Queen*: A Fresh Look at the Issues', *EM* 21 (1993), 44–62.

The musical scene in Act II is sung in fairyland ('a Prospect of Grotto's, Arbors, and delightful Walks') as Titania prepares for sleep. It falls into two halves: an entertainment put on for her by the fairies, and an extended masque-like lullaby. The first half uses the same fairy idiom as Act I—innocent, diatonic major harmonies, lightweight textures, dance forms and rhythms—though Purcell also paints an evocative musical landscape of the wood at night: a birdsong passage on a major-key *passacaglia* ground for two solo violins and continuo is followed by 'May the god of wit inspire', with its evocative double echoes of voices and trumpets and oboes. As Titania drifts towards sleep the key changes from C major to C minor, the dance forms largely disappear, and the music becomes rich and strange. Night sings an ethereal contrapuntal fantasia accompanied only by muted upper strings; Mystery follows with a langorous vocal gavotte; and Secrecy sings the famous 'One charming night gives more delight', a pseudo-ground with two recorders and continuo. Finally, Sleep enters with the command 'Hush, no more, be silent all' (Ex. 6.3). Roger North described the memorable effect of the 'lowd bass' and the 'full semibriefs pauses':

> Your fancy carrys the ratle of the instruments into those vacant spaces, and moreover, the subject and efficacy of the representation takes hold of your passion. And the equall measure of the pauses answers your expectation, that nothing can be greater and nobler than this was.[51]

This wonderful scene ends with the Dance for the Followers of Night, a creepy and dissonant double canon inspired by the one at the end of Locke's *Tempest* music.[52]

The music in Act III, an entertainment put on by Titania to charm Bottom, her new-found love, is not on this level, though it contains 'If love's a sweet passion', one of Purcell's loveliest minuet songs, 'Ye gentle spirits of the air', a spectacular da capo number, with a florid declamatory passage enclosing a minuet-like air, and the ever-popular rustic dialogue between Coridon and Mopsa. The masque in Act IV, also conjured up by Titania, is more substantial. The text ostensibly celebrates Oberon's birthday, though it largely consists of a Masque of the Four Seasons, introduced by Phoebus—an extended metaphor for the renewal of the

[51] Wilson, *Roger North*, 220–1; North remembered wrongly that the passage follows 'a disorder in swift music', and that it was sung by Hymen.

[52] Locke, *Dramatic Music*, 84–6.

Ex. 6.3. 'Hush, no more' from *The Fairy Queen* Z629/14, bars 1–11

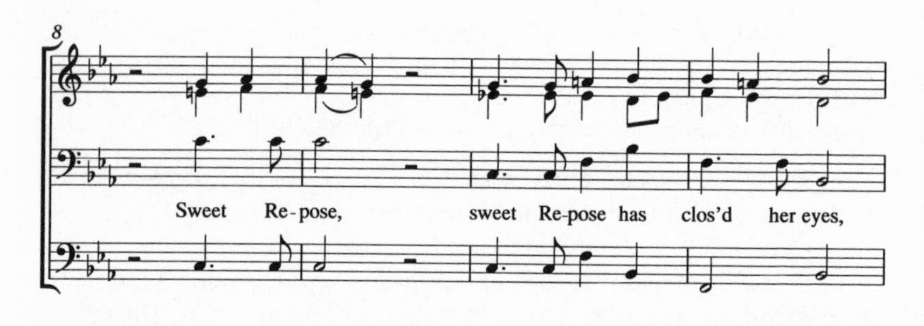

love between Titania and Oberon. Appropriately, Purcell draws heavily on the idiom of the court ode. It starts with a six-movement symphony similar in style, scoring, and structure to the one in 'Hail, bright Cecilia' Z328, composed later that year, and continues in similar vein, with ground-bass airs, and massive choruses. The airs of the seasons, allocated respectively to soprano, alto, tenor, and bass soloists, are surprisingly intimate, reflective pieces, the first three scored just with pairs of solo obbligato instruments. Winter's air, accompanied throughout by four-part strings, combines the anguished chromatic harmonies of the frost music in *King Arthur* with the grave contrapuntal textures of Night's air in Act II. The music of this scene is of the highest quality, but it must be said that it is no more dramatic than the average court ode.

The Fairy Queen ends with the usual masque, put on by Oberon to convince a sceptical Duke Theseus of the potency of magic, and incidentally celebrating love and marriage. Juno appears 'in a Machine drawn by Peacocks' and sings two contrasted numbers, the brilliant and Italianate 'Thrice happy lovers' and The Plaint, 'O let me weep'. The latter, a ground for soprano and an obbligato instrument over a freely varied *passacaglia* bass, is a puzzle. It was evidently added in 1693, for it is not in the Royal Academy score, and its text is not in the 1692 quarto. The main source is *Orpheus Britannicus*, i, pp. 194–9, where the obbligato part is allocated to a violin, but the simple 'vocal' quality of the part suggests a wind instrument, and its restricted range (*f*–*b*♭″) indicates a treble recorder, often associated by Purcell and his contemporaries with despairing love.[53] Such is the nature of criticism that the piece has been described as 'one of Purcell's finest [songs], pathos overwhelming the senses as in Dido's Lament' (Ian Spink), and 'in Purcell's best melancholy manner' (Bruce Wood and Andrew Pinnock), but as 'an over-indulgence; interminable, especially at the slow tempo at which it is usually performed' (Curtis Price).[54] My own sympathies lie with Price, and I wonder whether it is by someone else—such as Daniel Purcell, who was inclined to extend his brother's ideas beyond their natural life, and whose earliest datable music comes from 1693.

The rest of the masque does not have much more dramatic coherence than the Act V masque in *King Arthur*. Purcell made no effort to match the spectacular *chinoiserie* setting ('a transparent Prospect of a Chinese Garden, the Architecture, the Trees, the Plants, the Fruits, the Birds, the Beasts quite different to what we have in this part of the World') with exotic or even rustic music. 'Thus the gloomy world' is a superb da capo air in Purcell's best Italian manner, a C major duple-time ground with trumpet obbligato enclosing an A minor triple-time ground with two solo violins. But, sung by a Chinaman—whose nation is presumably

[53] Wood and Pinnock, '*The Fairy Queen*', 55, 62 suggest that the part is intended for an oboe.

[54] Spink, *English Song*, 229; Wood and Pinnock, '*The Fairy Queen*', 55; Price, *Henry Purcell*, 354.

meant to represent a state of rustic innocence—the piece could not be more incongrous.

Similar things could be said of the parent work. *The Fairy Queen* contains Purcell's most consistently inspired theatre music, yet much of it is devoted to episodes only tangentially connected to the drama, with few dramatic qualities of their own. Also, it cannot be said that the adaptor or adaptors provided Purcell with a wide range of dramatic opportunities, or that Purcell made an effort to provide the range of characters with appropriate musical idioms. He has been praised for finding a special tone for *The Fairy Queen*, a musical equivalent of Shakespeare's *faery*—Curtis Price diagnoses a partiality for augmented chords and minor dominants—but it does tend to be applied indiscriminately throughout the work. Nevertheless, *The Fairy Queen* can be a wonderful evening in the theatre, as a few sympathetic modern productions have shown.

The United Company did not put on any more full-blown semi-operas after *The Fairy Queen*, presumably because of the cost, and concentrated instead on plays with more modest musical and scenic requirements. The most original response to the change was D'Urfey's trilogy of plays from Cervantes, *The Comical History of Don Quixote*. Parts 1 and 2 were produced to great acclaim at Dorset Garden in May 1694 with songs mostly by Purcell and Eccles, published in two collections of *Songs to the New Play of Don Quixote*.[55] Part 3 followed in November 1695, and was a failure. Its songs were by a consortium, including Raphael Courteville, Thomas Morgan, and Samuel Akeroyde; Purcell's only surviving contribution is 'From rosy bowers', 'the last Song that Mr. Purcell Sett, it being in his Sickness', according to a note in *Orpheus Britannicus*, i, p. 90.

The *Don Quixote* plays are effectively semi-operas on the cheap. They require a good deal of music—Part 1, for instance, calls for eight songs and five dances, and probably would have had incidental music as well—and yet they require hardly any more scenic or musical resources than the average spoken play. None of the surviving songs needs more than three obbligato instruments (the recorders in Eccles's fine dialogue 'Sleep poor youth, sleep in peace', sung in Part 1, Act II, Scene ii), and some of them were given to speaking characters such as Marcella (played by Anne

[55] C. A. Price (ed.), *Don Quixote: The Music in the Three Plays of Thomas Durfey* (MLE, A-2; Tunbridge Wells, 1984) is a facs. edn. of all the surviving music for the plays.

Bracegirdle), Sancho Panza (Thomas Doggett), Cardenio (John Bowman), and Altisidora (Letitia Cross), which cut down the number of specialist singers needed.

It also had the effect of integrating the music into the action to an unusual extent. For example, Cardenio sings 'Let the dreadful engines of eternal will' (Part 1, Act IV, Scene i) on his first appearance 'in Ragged Cloaths, and in a Wild Posture', which efficiently explores his deranged mental state, induced by the faithlessness of his beloved.[56] Performers do not always appreciate that this famous song is a *comic* exploration of madness, but Purcell made the point unmistakably with ludicrous juxtapositions of declamatory passages in the grand manner and folk-like ballad tunes. Similarly, 'From rosy bowers' (Part 3, Act V, Scene i) does not have a tragic function: Altisidora is teasing the Don by feigning madness. Nevertheless, this great song has an enormous range, from skittish gaiety to bleak despair; in Curtis Price's apt words, it is 'the pure and unguarded expression of an artist who had no time left for artifice'.[57]

Another, less imaginative way that the United Company saved money in the 1690s was by reviving old tragedies, which meant that Purcell spent much of his time setting their ritual scenes. The earliest is probably the one in Act I, Scene iv of *Circe*, the only portion of the 1677 semi-opera reset by Purcell. Iphigenia, Priestess of Diana at the Scythian court, has attracted the love of King Thoas and Prince Ithacus; the situation leads Queen Circe to summon spirits to foretell the future. The music is mostly in major keys, and is rather bland and conventional, though there are a number of places where flattened sevenths push the music unexpectedly flatwards, suggesting more sinister undertones, and a priest sings a fine solo on a ground extolling the sweet smells and sounds of the ceremony; when it is taken up by the chorus the ground is abandoned and the melodic material thoroughly reworked, a pleasing touch. Most of the music could easily have been written around 1685, its traditional date, but the solo 'Lovers, who to their first embraces go', sung by one of Circe's women, has some expressive recitative as well as a passage of flying semiquavers in the voice and the bass, which suggests a rather later date; revivals of the play are recorded in June 1689 and November 1690.

[56] Price, *Henry Purcell*, 212–13. [57] Ibid. 219.

The incantation scene in Act III, Scene i of *Oedipus* Z583, a tragedy by Dryden and Lee from 1678, is more interesting, though it is more modestly scored, for ATB soloists, two violins, and continuo. Again, it survives only in undated manuscripts, but Charles Burney allocated it to 1692, and the play was reprinted then.[58] The blind seer Tiresias and two priests summon up the ghost of King Laius, hoping he will identify his murderer. They begin by addressing the 'sullen powers below', describing their unpleasant activities in a mixture of declamatory solos and tutti invocations. Then the first priest persuades Laius to rise by singing the famous ground-bass air 'Music for a while'; the scene ends with a tutti section commanding the ghost to appear.

'Music for a while' uses an arpeggiated ground with a simple rising bass in dialogue with an interlocking tenor, an effect first tried out in the odes of the early 1680s. The two strands creep upwards in an unpredictable mixture of chromatic and diatonic movement, portraying the eerie and inexorable rise of the dead king, while the melodic line soothes him with a series of gently descending phrases. As in a number of Purcell's late ground basses, the return to the home key after a series of modulations coincides with a return to the opening words and music, which gives it something of the character of a da capo aria (Ex. 6.4).

Purcell's music for a revival of *The Libertine* Z600, Shadwell's version of the Don Juan story, is normally assigned to 1692, though it seems to be a late work. The song 'To arms, heroic prince', an Italianate two-section air for soprano, trumpet, and continuo, is said in *Deliciae musicae*, I/ii (July 1695) to have been sung in the play by 'the Boy'. If, as seems likely, this is Jemmy Bowen, then it cannot date from before the spring of 1695, when he began his theatre career.[59] Of course, the song, not called for by the text, could have been sung by Bowen between the acts of a later revival. But another clue is provided by the 'flourish' that begins the Act V music. It is essentially the same as the March Z860/1, played at Queen Mary's funeral on 5 March 1695. Zimmerman assumed that the theatre version came first, but Margaret Laurie argued that it was the other way round, and Bruce Wood has pointed out that the bass of Z860/1 does not go below *c*, whereas in Z600/2a it goes down to *G* or (in some sources) *E*♭, presumably because four flat trumpets were not avail-

[58] Burney, *General History*, ii. 390.

[59] Price, *Henry Purcell*, 114.

Ex. 6.4. 'Music for a while' from *Oedipus* Z583/2, bars 22–31

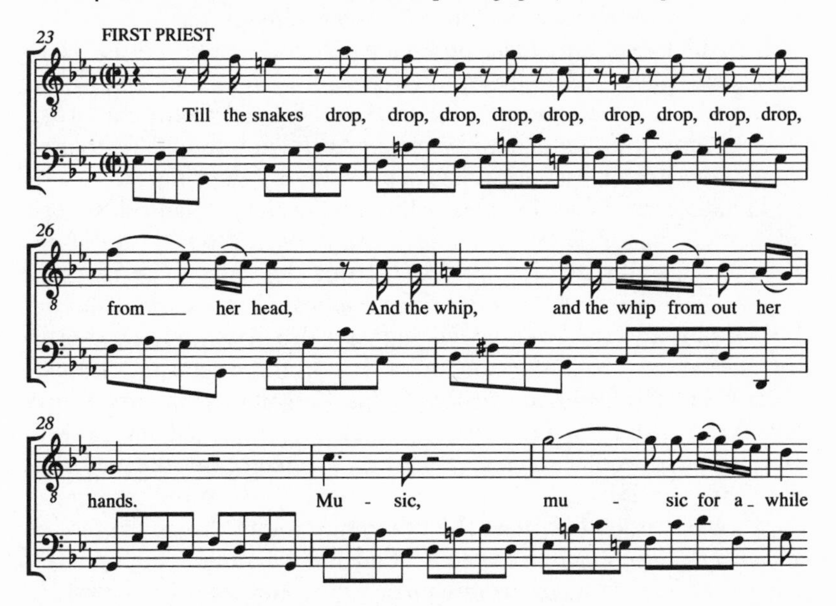

able in the theatre, and the part had to be played on bass violins or bassoons.[60] Thus the theatre piece is apparently an adaptation of the funeral march, which means that *The Libertine* probably dates from the summer or autumn of 1695.

We have Purcell's music for two scenes. The first, an episode of rustic merriment in Act IV set in 'a delightful grove', consists of the famous song 'Nymphs and shepherds, come away' with its prelude for four-part strings ('A Symphony of Rustick Musick'), and an extended chorus. The prelude seems to be an imitation of a village band, with three-chord harmonies and aimlessly busy upper parts imitating a collective improvisation. The song starts with the same idea, and sounds artless to the casual listener. In fact, it is a highly sophisticated piece, with an Italianate harmonic plan taking the music smoothly from G major through a central section in E minor and A minor to an abbreviated da capo.

The other episode, at the end of the play, is a 'Song of Devils', scored apparently just for STB soloists (despite the layout of the score in *Works I*, 20), strings, and continuo. Given that Don Juan and his cronies have just been offered glasses of blood, and are

[60] Zimmerman, *Purcell: Catalogue*, 275; id., *Purcell: Life and Times*, 212, 242–3; Laurie, 'Purcell's Stage Works', 214–15; B. Wood, 'First Performance of Purcell's Funeral Music'.

about to be dragged down to Hell, the music is surprisingly restrained and ecclesiastical in character, though there are some wonderful twists and turns in the vocal line at the words 'Here they shall weep and shall unpitied groan, | Here they shall howl and make eternal moan.' Restoration devils, from the 1674 *Tempest* onwards, tended to be given fairly simple and restrained music; the horror was presumably provided by the singers.

The date of Purcell's music for a revival of Shadwell's 1678 adaptation of *Timon of Athens* Z632 is also in question. Zimmerman allocates it to '1694(?)', but the song 'The cares of lovers', like 'To arms, heroic prince', appeared in *Deliciae musicae*, I/ii as sung by 'the boy', which suggests that it too was written in the spring of 1695. On the other hand, the D major trumpet overture is a transposed version of the C major symphony to the ode 'Who can from joy refrain?' Z342/1, performed on 24 July 1695, which would suggest a first performance in the autumn; there are several cases of Purcell borrowing overtures from odes for theatre works, but none the other way round. Some sources of *Timon of Athens* include a suite of incidental music, but it is by James Paisible, to judge from concordances and its musical style.[61] Purcell seems to have written only one instrumental piece specially for the play, the wonderful G minor 'Curtain Tune' on a chromatic ground bass. It apparently duplicates the overture, and Curtis Price has argued that it is mislabelled, and was actually played in Act IV, Scene iii while Timon throws stones at his tormentors.[62]

Purcell's main contribution to the play is the Masque of Cupid and Bacchus in Act II, Scene ii, performed during a banquet thrown by Timon for his opportunistic cronies; it serves as a warning to Timon not to prefer the flighty Melissa to the faithful Evandra. It is a symmetrically patterned debate on the merits of love and wine. There are opening statements by their supporters, appearances by Cupid and Bacchus, and more contributions by the supporters; the masque ends in jolly agreement. It is clear from the music where Purcell's sympathies lay. Wine is portrayed as a limited, unvaried attraction, always in B flat major, always with oboes, and always in rather stolid dance rhythms. Love, by contrast, is varied—in key, scoring, and musical style. Moreover, he gets the best pieces: the charming F major opening duet 'Hark!

[61] Price, *Henry Purcell*, 94. [62] Ibid. 94–6.

how the songsters of the grove' with two recorders (Purcell takes his cue from the stage direction 'A Symphony of Pipes imitating the chirping of Birds') over a bouncing pseudo-ground; the heartfelt F minor trio 'But ah! how much are our delights more dear!', which hints that pain is the consequence of love; the beautiful D minor minuet chorus 'Who can resist such mighty charms?'; and 'The cares of lovers', a dazzling G minor recitative. The only puzzle is that the work ends so tamely, with a jig-like number in B flat, the Bacchic key. But, as Price suggests, that may be deliberate, hinting that Melissa/wine will triumph after all.

The Indian Queen Z630 is Purcell's greatest late theatre work. The date of its first performance has long been a matter of controversy. Curtis Price has argued that he wrote at least some of it before the breakup of the United Company in March 1695, which forced it to be shelved.[63] For how long is not known, but it may have been performed in June 1695 without the Act V masque, or it may not have been seen until after his death; the masque was set by Daniel Purcell, who perhaps would not have been given the task while his brother was alive. The plot involves accepting that the Inca and Aztec empires in ancient Peru and Mexico are adjacent and at war. It concerns the Aztec queen Zempoalla, and her unrequited love for Montezuma, a warrior of unknown origin; he leads the Incas to victory, only to have his suit for Princess Orazia rejected, whereupon he changes sides.

Purcell's music is concentrated in four main episodes. The first, the prologue, is more interesting for the fact that it was set to music than for the music itself. Instead of the normal topical speech an Indian couple sing a series of graceful airs, alluding to the events of the play, and foretelling the eventual Spanish conquest. The Masque of Fame and Envy praising Zempoalla comes in Act II, Scene ii, though Andrew Pinnock has recently argued that it more properly belongs to Act III.[64] It starts with the symphony from 'Come, ye sons of art, away' Z323/1, transposed into C major. Like the masque in *Timon of Athens*, the protagonists are characterized by key, scoring, and musical style. Fame, a heroic tenor, is given swinging C major tunes in the trumpet idiom; the first, 'We come to sing Zempoalla's story', is literally a trumpet tune, as is revealed when the chorus and orchestra take it up.

[63] Price, *Henry Purcell*, 126–31 summarizes the problems; see also A. Pinnock, 'Play into Opera: Purcell's *The Indian Queen*', *EM* 18 (1990), 3–21.
[64] Ibid. 17–18.

Ex. 6.5. 'What flatt'ring noise is this' from *The Indian Queen* Z630/7, bars 1–9

Envy, a bass, is given sinister C minor music, accompanied by two solo violins, and has two accomplices in 'What flatt'ring noise is this? | At which my snakes all hiss?' who provide the snake noises—an absurdly simple masterstroke which, once heard, can never be forgotten (Ex. 6.5).

The greatest music in *The Indian Queen* comes in the conjuring scene in Act III, Scene ii. Zempoalla's unrequited love leads her to consult the magician Ismeron, who summons up the God of Dreams to reveal her fate. 'You twice ten hundred deities' is the most awe-inspiring of Purcell's musical conjuring tricks. It is in four sections, each corresponding to a stage in the action. The magician starts in conventional recitative, addressing all the gods 'to whom we daily sacrifice'. Then he singles out the God of Dreams, and the music becomes more and more laden with dissonance, reaching an anguished chromatic conclusion at the words 'what strange fate must on her dismal vision wait'. Ismeron, Purcell tells us, already knows her fate. In the air he recites a litany of charms, each one dispatched by a sinister, jerky figure in the violins. The voice mostly alternates with the violins, so that the singer can extract every ounce of vocal colour and menace from his voice without having to compete with them. Suddenly, the second violin begins a slow chromatic ascent, and we know that the charms have worked: the god is starting to rise through the trapdoor. Ismeron joins in, ascending an octave in a largely chromatic line, and ending with a shattering, dissonant climax at the words 'open thy unwilling eyes' (Ex. 6.6). Then the exhausted magician relaxes as Purcell illustrates the 'bubbling streams' in graceful falling patterns, harmonized in sweet sixths and thirds.

Inevitably, what follows is an anticlimax, though it begins with a fine, stately symphony for two oboes and continuo, presumably played as the god moves slowly to the front of the stage. His tripartite air with oboe obbligato, 'Seek not to know', is disappointing, though its modern, courtly style—reminiscent of Purcell's younger contemporaries such as William Croft or John Eccles—is perhaps intended to portray the god as a diplomat, telling the queen as little as possible as politely as possible. In the second part of the scene Ismeron conjures ariel spirits to console Zempoalla. A superb trumpet overture is followed by the ground 'Ah! how happy are we', and 'We, the spirits of the air', another delicious gavotte in rondeau form for two sopranos with chorus. Purcell's last contribution to *The Indian Queen*, the Sacrifice Scene

Ex. 6.6. 'You twice ten hundred deities' from *The Indian Queen* Z630/13, bars 58–74

in Act V, is short and largely conventional, though it comes to life as the High Priest asks the company 'Is all prepared?' in music reminiscent of that given to the Sorceress in *Dido*.

The Indian Queen, like *King Arthur*, is beset by editorial problems, despite the fact that we have a manuscript, GB-Lbl, Add. MS 31449, that has the text interleaved with the music. Unfortunately, it is none too accurate, and there are a number of ambiguities. Should Ismeron speak 'You twice ten hundred deities' before singing it, as the manuscript suggests? No, argues Andrew Pinnock, surely correctly.[65] Should 'I attempt from love's sickness to fly' be sung at the end of the Act III music, where the music copyist of Add. MS 31449 placed it (followed in *Works I*, 19), or should it come at the beginning of that scene, where there is the stage direction 'Ismeron asleep in the scene | Song here: Enter Zempolla'.[66] It certainly makes sense there, for the song, yet another beautiful minuet in rondeau form, represents her feelings of longing and confusion at that point in the play. There are similar problems with the placing of several other movements, and in general *The Indian Queen* illustrates the need for scholarly editions of these wonderful works that bring the words and the music together.

Purcell's last major stage work was probably *Bonduca* Z574, as adapted for a revival in the autumn of 1695. Fletcher's play chronicles the struggles of the 'British heroine' (Boadicea or Boudicca to us) against the Romans. Most of the music comes in another temple scene, in Act III, Scene ii, with Druids praying to the gods for a victory. Little need be said about it, except that it contains a fine bass solo, 'Hear, ye gods of Britain', with some hair-raising harmonies in the four-part string accompaniment. After the oracle has delivered an ambiguous message, 'Much [blood] will be spill'd', they prepare for war with 'To arms, your ensigns straight display' and 'Britons, strike home'.

In the subsequent battles the Britons are defeated, so there is an irony in the fact that 'Britons, strike home', a simple minuet-like trumpet tune, became a patriotic favourite, and remained so for much of the next century. Sheridan, for instance, called for it with 'Rule, Britannia' and 'See, the conquering hero comes' in the masque with model boats that ends *The Critic*. The finest piece in *Bonduca* is the song 'O lead me to some peaceful gloom', sung by

[65] Pinnock, 'Play into Opera', 13–14.

[66] Price, *Henry Purcell*, 140; Pinnock, 'Play into Opera', 8–9.

Princess Bonvica in Act V as she prepares to die. It is a beautiful example of the double-barrelled song, with a wide-ranging declamatory passage balanced by a vocal minuet. As in many of Purcell's late songs, much of the power and strength of the music comes from the bass. It suggests the girl's approaching fate with the simplest of means at the opening (Ex. 6.7), and it conducts a dialogue on equal terms with the voice in the second section.

Ex. 6.7. 'O lead me to some peaceful gloom' from *Bonduca* Z574/17, bars 1–9

In a book of this length it is not possible to discuss all of Purcell's smaller theatre works, but something must be said about his theatre suites. The publication *A Collection of Ayres, Compos'd for the Theatre, and upon other Occasions* was published in 1697, and contains thirteen suites written for plays between June 1690 and the autumn of 1695—that is, between *Dioclesian* and *Bonduca*. As far as is known, it contains nothing but theatre music, despite the mention of 'other occasions', and it contains all his theatre suites, omitting Grabu's for *The Double Marriage*, Paisible's for *Timon of Athens*, and the incomplete one for Robert Gould's *The*

Rival Sisters Z609 (November 1695), which may be by the violinist John Ridgley, apart from the overture, which was borrowed from 'Love's goddess sure was blind' Z331/1.[67] *Ayres for the Theatre*, as it was known for short, was the first publication devoted entirely to theatre suites, and was the model for *Harmonia Anglicana*, the series of more than fifty theatre suites published by Walsh and Hare in the first decade of the eighteenth century.

In *Ayres for the Theatre* the movements of the suites were shuffled to make satisfactory sequences for use in concerts, often ordered by key. The one for *The Indian Queen*, for instance, starts in C minor with the overture (no. 3), and proceeds by way of C major (nos. 4a and 6), F major (nos. 1a, 2d, 1b, 2a), D minor (no. 22), A minor (no. 17), to A major (no. 18). Two movements (nos. 6 and 17) are instrumental versions of vocal numbers, 'I come to sing great Zempoalla's story' and 'We, the spirits of the air', and were probably included as 'hit songs' from the show; they were not necessarily used in the production as preludes to the songs. The collection is the primary source for most of the suites for ordinary spoken plays, so in most cases we do not know what the original order of the movements was. There are secondary manuscripts of *Abdelazer* Z570, *Amphitryon* Z572, *Distressed Innocence* Z577, *The Gordian Knot Unty'd* Z597, and *The Virtuous Wife* Z611 with the movements in theatre order, the overture placed after the second music, though it is not clear in all cases what authority they have.[68]

Ayres for the Theatre has been maligned as an inaccurate source that wantonly omits essential wind and brass parts. In fact, most of the suites are complete as they stand in four parts (there is no continuo part, no figuring in the bass, and little sign that they were played with continuo, in the theatre at least), and the texts are no more inaccurate than in most consort collections of the time. True, there are movements with missing parts in the suites from the semi-operas; the worst case is the overture to *The Fairy Queen* Z629/3, printed without trumpets. But, in general, the selections were done with skill and discrimination, and make excellent concert material. Purcell evidently took the writing of his theatre suites seriously, and Curtis Price has rightly remarked that 'one can hardly find a single lacklustre piece' in *Ayres for the Theatre*.[69]

[67] Zimmerman, *Purcell: Catalogue*, 288.

[68] Ibid. 240–1, 252–3, 271–3, 289–90; Price, *Henry Purcell*, 150–2.

[69] Price, *Henry Purcell*, 66.

Yet there are few signs that Purcell tried to convey the mood of the play in his suites, beyond a tendency to preface tragedies such as *Abdelazer* and *Distressed Innocence* with overtures wholly or partly in the minor. *Bonduca*, Price has aruged, is an exception: the overture, scored for trumpet and strings (the trumpet part is not in *Works I*, 16, and only survives in manuscript), prefigures the denouement of the tragedy by turning from the C major fugue to a wild chromatic close in C minor.[70]

The overtures to Purcell's suites are often characterized as 'French', but a number of their first sections are cast in patterns of interlocking semiquavers (*Abdelazer*, *The Virtuous Wife*), or as an almand-like passage (*Amphitryon*, *Distressed Innocence*), rather than in the dotted rhythms of the French type. The fugue in the *Amphitryon* overture is a rare example in Purcell of a common type in Lully, with a subject in flowing quavers that enters in sequence from first violin to bass (the second violin comes in a bar too late in *Works I*, 16). Some of the fugues are actually more Italian than French: the duple-time one in *Abdelazer* with its striding chromatic theme would not be out of place in an Italian-style sonata, and the same can be said of the closely argued fugue in *The Virtuous Wife*. These fine pieces are virtually monothematic, but Purcell also favoured the type in which the first theme is replaced halfway through by a second, as in *The Double Dealer* Z592/1 or *The Married Beau* Z603/1, a device that enabled him to expand the dimensions of his fugues without diluting their contrapuntal energy in episodes.

The rest of the suites are, if anything, less French than the overtures. There is no shortage of elegant minuets and minuet-like airs, but the most common dance is the hornpipe (fifteen examples in the nine suites for ordinary spoken plays), and throughout there is a preponderance of breezy, tuneful airs, some of which achieved the status of popular tunes. Successive editions of *The Dancing Master* until the 1720s contain the hornpipe from *Abdelazer* Z570/8 (not to be confused with the rondeau in hornpipe rhythm Z570/2, made famous by Benjamin Britten), one of the hornpipes from *Bonduca* Z574/7, and the hornpipe from *Dioclesian* Z627/4 under the titles 'The hole in the wall', 'Westminster Hall', and 'The siege of Limerick'; Z627/4 also circulated as the song 'O how happy's he who from bus'ness free'

[70] Price, *Henry Purcell*, 118–19.

Z403, the words added to the tune by the actor William Mountfort.[71] Posthumous fame of this sort is telling evidence that the English musical world took *Ayres for the Theatre* and its delightful suites to its collective heart, and did not easily relinquish it.

[71] J. Barlow (ed.), *The Complete Country Dance Tunes from Playford's Dancing Master (1651–ca. 1728)* (London, 1985), nos. 343, 377, 392; C. L. Day and E. Boswell Murrie, *English Song-Books 1651–1702, a Bibliography* (London, 1940), no. 2474.

BIBLIOGRAPHY

Books and Articles

ADAMS, M., 'More on Dating *Dido*', *EM* 21 (1993), 510.

ALLSOP, P., 'The Role of the Stringed Bass as a Continuo Instrument in Italian Seventeenth-Century Instrumental Music', *Chelys*, 8 (1978–9), 31–7.

—— 'Problems of Ascription in the Roman *Simfonia* of the Late Seventeenth Century: Colista and Lonati', *MR* 50 (1989), 34–44.

—— *The Italian 'Trio' Sonata from its Origins until Corelli* (Oxford, 1992).

ARNOLD, F. T., *The Art of Accompaniment from a Thorough-Bass* (London, 1931; repr. 1965).

ASHBEE, A. (ed.), *Records of English Court Music* [*RECM*], i: *1660–1685* (Snodland, 1986).

—— *RECM* ii: *1685–1714* (Snodland, 1987).

—— *RECM* v: *1625–1714* (Aldershot, 1991).

—— *The Harmonious Musick of John Jenkins*, i: *The Fantasias for Viols* (Surbiton, 1992).

AUBREY, J., *Brief Lives*, ed. R. Barber (2nd edn., Woodbridge, 1982).

BALDWIN, D., *The Chapel Royal Ancient and Modern* (London, 1990).

BALDWIN, O., and WILSON, T., 'Musick Advanced and Vindicated', *MT* 111 (1970), 148–50.

—— '"Who Can from Joy Refraine?": Purcell's Birthday Song for the Duke of Gloucester', *MT* 122 (1981), 596–9.

[BANISTER I, J.], *Musick; or, A Parley of Instruments, the First Part* (London, 1676).

BANISTER II, J., *The Sprightly Companion* (London, 1695; repr. 1984).

BARTLETT, C., Notes to the Consort of Musicke's recording, *Henry Lawes: Sitting by the Streams, Psalmes, Ayres, and Dialogues*, Hyperion A66135 (1984).

BOALCH, D. H., *Makers of the Harpsichord and Clavichord 1440–1840* (2nd edn., Oxford, 1974).

BRENNECKE, E., 'Dryden's Odes and Draghi's Music', *Proceedings of the Modern Language Association of America*, 49 (1934), 1–36.

BROWNING, A., 'Purcell's "Stairre Case Overture"', *MT* 121 (1980), 768–9.

BUKOFZER, M. (ed.), *Giovanni Coperario: Rules How to Compose* (San Marino, Calif., 1952).

BUMPUS, J. S., *A History of English Cathedral Music 1559–1889* (London, 1889).

Burden, M. (ed.), *Performing the Music of Henry Purcell* (Oxford, forthcoming).
Burney, C., *A General History of Music* (London, 1776–89), ed. F. Mercer (London, 1935; repr. 1957).
Butler, C., *The Principles of Musik in Singing and Setting* (London, 1636; repr. 1970).
Buttrey, J., 'The Evolution of English Opera between 1656 and 1695: A Reinvestigation' (Ph.D. thesis, Cambridge, 1967).
—— 'Dating Purcell's *Dido and Aeneas*', *PRMA* 94 (1967–8), 51–62.
Campbell, M., *Henry Purcell: Glory of his Age* (London, 1993).
Catalogue of the Musical Library, etc. of the Late Rev. Samuel Picart, 10 March 1848, GB-Lbl, S.C.P.6(1).
Chan, M., 'Drolls, Drolleries, and Mid-Seventeenth-Century Dramatic Music in England', *RMARC* 15 (1979), 117–73.
Chappell, W., *The Ballad Literature and Popular Music of the Olden Time* (London, 1859; repr. 1965).
Charteris, R., 'Some Manuscript Discoveries of Henry Purcell and his Contemporaries in the Newberry Library, Chicago', *Notes*, 37 (1980), 7–13.
—— *A Catalogue of the Printed Books on Music, Printed Music, and Music Manuscripts in Archbishop Marsh's Library, Dublin* (Clifden, 1982).
Cholij, I., and Price, C. A., 'Purcell's Bass Sorceress', *MT* 127 (1986), 615–18.
Cooper, B., 'Keyboard Music', in Spink (ed.), *Seventeenth Century*, 341–66.
Cowley, A., *Pindarique Odes, Written in Imitation of the Stile and Manner of the Odes of Pindar* (London, 1656).
Cox, G., *Organ Music in Restoration England: A Study of Sources, Styles, and Influences* (New York and London, 1989).
CSPD, James II, February–December 1685, ed. E. K. Timings (London, 1960).
Cummings, W. H., *Henry Purcell 1658–1695* (2nd edn., London, 1911).
—— 'The Mutilation of a Masterpiece', *PMA* 30 (1903–4), 113–27.
Dart, T., 'Purcell's Chamber Music', *PRMA* 85 (1958–9), 81–93.
—— 'Purcell and Bull', *MT* 104 (1963), 30–1.
Dawe, D., *Organists of the City of London 1666–1850* (Padstow, 1983).
Day, C. L., and Boswell Murrie, E., *English Song Books 1651–1702, a Bibliography* (London, 1940).
Dearnley, C., *English Church Music 1650–1750* (London, 1970).
Dennison, P., 'Purcell: The Early Church Music', in Sternfeld *et al.* (eds.), *Essays on Opera and English Music*, 44–61.
—— *Pelham Humfrey* (Oxford, 1986).
Downes, J., *Roscius Anglicanus* (London, 1708), ed. J. Milhous and R. D. Hume (London, 1987).

DOWNEY, P., 'What Samuel Pepys Heard on 3 February 1661: English Trumpet Style under the Later Stuart Monarchs', *EM* 18 (1990), 417–28.

DRYDEN, J., *Works*, 15: *Plays: Albion and Albanius, Don Sebastian, Amphitryon*, ed. E. Miner, G. R. Guffey, and F. B. Zimmerman (Berkeley, Los Angeles, and London, 1976).

DUCKLES, V., 'Florid Embellishment in English Song of the Late 16th Century and Early 17th Centuries', *AnnM* 5 (1957), 329–45.

EVELYN, J., *The Diary of John Evelyn*, ed. E. S. de Beer, 6 vols. (London, 1955).

EWARD, S., *No Fine but a Glass of Fine Wine: Cathedral Life at Gloucester in Stuart Times* (Wilton, 1985).

FIELD, C. D. S., and TILMOUTH, M., 'Consort Music II: From 1660', in Spink (ed.), *Seventeenth Century*, 245–81.

FORD, R., 'Osborn MS 515: A Guardbook of Restoration Instrumental Music', *FAM* 30 (1983), 174–84.

—— 'Purcell as his Own Editor: The Funeral Sentences', *Journal of Musicological Research*, 7 (1986), 47–67.

FORTUNE, N., 'Purcell: The Domestic Sacred Music', in Sternfeld *et al.* (eds.), *Essays on Opera and English Music*, 62–78.

—— and ZIMMERMAN, F. B., 'Purcell's Autographs', in Holst (ed.), *Purcell: Essays on His Music*, 106–21.

FREEMAN, A., 'Organs Built for the Royal Palace at Whitehall', *MT* 52 (1911), 720–1.

GIANNINI, T., 'A Letter from Louis Rousselet, Eighteenth-Century French Oboist at the Royal Opera in England', *Newsletter of the American Musical Instrument Society*, 16/2 (June 1987), 10–11.

GILDON, C., *The Life of Mr. Thomas Betterton* (London, 1710; repr. 1970).

GOLDIE, M., 'The Earliest Notice of Purcell's *Dido and Aeneas*', *EM* 20 (1992), 392–400.

GRAY, T., *Correspondence*, i: *1734–1755*, ed. P. Toynbee and L. Whibley (Oxford, 1935).

GWYNN, D., 'The English Organ in Purcell's Lifetime', in Burden (ed.), *Performing the Music of Purcell*.

HAND, C., *John Taverner: His Life and Music* (London, 1978).

HARDING, R., *A Thematic Catalogue of the Works of Matthew Locke* (Oxford, 1971).

HARRIS, E. T., *Henry Purcell's Dido and Aeneas* (Oxford, 1987).

HART, E. F., 'The Restoration Catch', *ML* 34 (1953), 288–305.

HAWKINS, J., *A General History of the Science and Practice of Music* (London, 1776; repr. 1853 and 1963).

HAYNES, B., 'Johann Sebastian Bach's Pitch Standards: The Woodwind Perspective', *JAMIS* 11 (1985), 55–114.

HOGWOOD, C., 'Thomas Tudway's History of Music', in Hogwood and Luckett (eds.), *Music in Eighteenth-Century England*, 19–47.
—— and LUCKETT, R. (eds.), *Music in Eighteenth-Century England: Essays in Memory of Charles Cudworth* (Cambridge, 1983).
HOLMAN, P., 'Suites by Jenkins Rediscovered', *EM* 6 (1978), 23–35, and the correspondence ibid. 481–3.
—— 'Thomas Baltzar (?1631–1663), the "Incomperable Lubicer on the Violin"', *Chelys*, 13 (1984), 3–38.
—— 'A New Source of Restoration Keyboard Music', *RMARC* 20 (1986–7), 53–7.
—— 'Bartholomew Isaack and "Mr Isaack" of Eton: A Confusing Tale of Restoration Musicians', *MT* 128 (1987), 381–5.
—— Review of *Dido and Aeneas*, ed. E. Harris, *ML* 71 (1990), 617–20.
—— Response to Downey, 'What Samuel Pepys Heard', *EM* 19 (1991), 443.
—— '"An Addicion of Wyer Stringes beside the Ordenary Stringes": The Origin of the Baryton', in J. Paynter, R. Orton, P. Seymour, and T. Howell (eds.), *Companion to Contemporary Musical Thought* (London and New York, 1992), 1092–1115.
—— *Four and Twenty Fiddlers: The Violin at the English Court 1540–1690* (Oxford, 1993).
—— 'Original Sets of Parts for Restoration Concerted Music at Oxford', in Burden (ed.), *Performing the Music of Purcell*.
—— 'Henry Purcell and Daniel Roseingrave: A New Autograph', in Price (ed.), *Purcell Studies*.
HOLST, I. (ed.), *Henry Purcell 1659–1695: Essays on his Music* (London, 1959).
HOTSON, L., *The Commonwealth and Restoration Stage* (New York, 1962).
HUSK, W. H., *An Account of the Musical Celebrations on St. Cecilia's Day* (London, 1857).
ILLING, R., *Henry Purcell: Sonata in G Minor for Violin and Continuo: An Account of its Survival from both the Historical and Technical Points of View* (Flinders University, South Australia, 1975).
JOHNSTONE, H. D., 'English Solo Song, *c.*1710–1760', *PRMA* 95 (1968–9), 67–80.
JONES, E. H., *The Performance of English Song, 1610–1670* (New York and London, 1989).
KLAKOWICH, R., 'Harpsichord Music by Purcell and Clarke in Los Angeles', *Journal of Musicology*, 4 (1985–6), 171–90.
—— '"Scocca pur": Genesis of an English Ground', *JRMA* 116 (1991), 63–77.
KRUMMEL, D., *English Music Printing 1553–1700* (London, 1975).
LAFONTAINE, H. C. DE (ed.), *The King's Musick: A Transcript of Records Relating to Music and Musicians (1460–1700)* (London, 1909; repr. 1973).

LASOCKI, D., 'The Detroit Recorder Manuscript (England, *c.*1700)', *American Recorder*, 23/3 (Aug. 1982), 95–102.
—— 'The Anglo-Venetian Bassano Family as Instrument Makers and Repairers', *GSJ* 38 (1985), 112–32.
—— 'The French Hautboy in England, 1673–1730', *EM* 16 (1988), 339–57.
LAURIE, A. M., 'Purcell's Stage Works' (Ph.D. thesis, Cambridge, 1961).
—— 'Did Purcell Set *The Tempest*?', *PRMA* 90 (1963–4), 43–57.
—— 'Purcell's Extended Solo Songs', *MT* 125 (1984), 19–25.
—— 'Music for the Stage II: From 1650', in Spink (ed.), *Seventeenth Century*, 306–40.
LAWRENCE, W. J., 'Foreign Singers and Musicians at the Court of Charles II', *MQ* 9 (1923), 217–25.
LEFKOWITZ, M., *William Lawes* (London, 1960).
LE HURAY, P., *Music and the Reformation in England, 1549–1660* (London, 1967).
LOCKE, M., *The Present Practice of Musick Vindicated* (London, 1673, repr. 1974).
LOVE, H., 'The Wreck of the Gloucester', *MT* 125 (1984), 194–5.
LOWE, E., *A Short Direction for the Performance of the Cathedrall Service* (London, 1661).
—— *A Review of a Short Direction* (London, 1664).
LUCKETT, R., '"Or Rather our Musical Shakespeare": Charles Burney's Purcell', in Hogwood and Luckett (eds.), *Music in Eighteenth-Century England*, 59–77.
—— 'A New Source for *Venus and Adonis*', *MT* 130 (1989), 76–80.
LUTTRELL, N., *A Brief Historical Relation of State Affairs from September 1678 to April 1714* (Oxford, 1857; repr. 1969).
MABBETT, M., 'Italian Musicians in Restoration England (1660–90)', *ML* 67 (1986), 237–47.
MACE, T., *Musick's Monument* (London, 1676; repr. 1958).
MCGUINNESS, R., 'The Ground-Bass in the English Court Ode', *ML* 51 (1970), 118–40, 265–78.
—— *English Court Odes 1660–1820* (Oxford, 1971).
MAGALOTTI, L., *Travels of Cosmo the Third, Grand Duke of Tuscany through England, during the Reign of King Charles the Second (1669)* (London, 1821).
MANNING, R., 'Revisions and Reworkings in Purcell's Anthems', *Soundings*, 9 (1982), 29–37.
MILHOUS, J., 'The Date and Import of the Financial Plan for a United Theatre Company in P.R.O. LC 7/3', *Maske und Kothurn*, 21 (1975), 81–8.
—— and PRICE, C. A., 'Harpsichords in the London Theatres, 1697–1715', *EM* 18 (1990), 38–46.

MONSON, C., *Voices and Viols in England, 1600–1650: The Sources and their Music* (Ann Arbor, 1982).
MORLEY, T., *A Plaine and Easie Introduction to Practicall Musicke* (London, 1597; repr. 1971).
MORRISON, R., 'Purcell's Notebook Revealed', *The Times* (17 Nov. 1993), 35.
MULLER, J., *Words and Music in Henry Purcell's First Semi-Opera, Dioclesian* (Lewiston, NY, Queenston, Ontario, and Lampeter, Dyfed, 1990).
Nanki Music Library [Summary Catalogue] (Tokyo, 1967).
NICHOLS, D., 'Edward Finch and the Associated Influential Composers and their Music' (BA diss, Colchester Institute, 1989).
NICOLL, A., *A History of English Drama*, i: *Restoration Drama, 1660–1700* (4th edn., Cambridge, 1952); ii: *Early Eighteenth Century Drama* (3rd edn., Cambridge, 1952).
PEPYS, S., *Letters and the Second Diary of Samuel Pepys*, ed. R. G. Howarth (London, 1932).
—— *The Diary of Samuel Pepys*, ed. R. Latham and W. Matthews, 11 vols. (London, 1970–83).
PINNOCK, A., 'Play into Opera: Purcell's *The Indian Queen*', *EM* 18 (1990), 3–21.
—— and WOOD, B., 'A Counterblast on English Trumpets', *EM* 19 (1991), 436–43.
—— '"Unscarr'd by Turning Times"?: The Dating of Purcell's *Dido and Aeneas*', *EM* 20 (1992), 372–90.
—— '*The Fairy Queen*: A Fresh Look at the Issues', *EM* 21 (1993), 44–62.
—— '"Singin' in the Rain": Yet More on Dating *Dido*', *EM* 22 (1994), 365–7.
PINTO, D., 'William Lawes's Music for Viol Consort', *EM* 6 (1978), 12–24.
—— 'The Fantasy Manner', *Chelys*, 10 (1981), 17–28.
—— 'The Music of the Hattons', *RMARC* 23 (1990), 79–108.
PLAYFORD, J., *A Breif (Brief) Introduction to the Skill of Musick* (London, 1654; 1655; 1666; 7th edn., 1674; repr. 1966; 12th edn., 1694; repr. 1972).
PRICE, C. A., *Music in the Restoration Theatre* (Ann Arbor, Mich., 1979).
—— *Henry Purcell and the London Stage* (Cambridge, 1984).
—— '*Dido and Aeneas:* Questions of Style and Evidence', *EM* 22 (1994), 115–25.
—— (ed.), *Purcell Studies* (Cambridge, forthcoming).
PRIOR, M., *Poems on Several Occasions*, ed. A. R. Waller (Cambridge, 1905).
PROTHEROE, J., 'Not so much an Opera . . . a Restoration Problem Examined', *MT* 106 (1965), 666–8.

RIMBAULT, E. F. (ed.), *The Old Cheque-Book or Book of Remembrance of the Chapel Royal from 1561 to 1744* (London, 1872; repr. 1966).
ROSE, G., 'Pietro Reggio—a Wandering Musician', *ML* 46 (1965), 207–16.
—— 'Purcell, Michaelangelo Rossi, and J. S. Bach: Problems of Authorship', *AcM* 40 (1968), 203–19.
RUFF, L., 'Thomas Salmon's "Essay to the Advancement of Musick"', *The Consort*, 21 (1964), 266–78.
SADIE, S. (ed.), *The New Grove Dictionary of Music and Musicians* [*Grove 6*], 20 vols. (London, 1980).
SAINTSBURY, G. (ed.), *Minor Poets of the Caroline Period* (London, 1921).
SAVAGE, R., 'The Shakespeare–Purcell *Fairy Queen*', *EM* 1 (1973), 200–21.
SCHOLES, P., *The Puritans and Music in England and New England* (Oxford, 1934; repr. 1969).
[SHADWELL, T.], *The Tempest; or, The Enchanted Island* (London, 1674).
SHAW, H. W., 'A Collection of Musical Manuscripts in the Autograph of Henry Purcell and Other English Composers *c.* 1665–85', *The Library*, 14 (1959), 126–31.
—— 'A Contemporary Source of English Music of the Purcellian Period', *AcM* 31 (1959), 38–44.
—— 'A Cambridge Manuscript from the English Chapel Royal', *ML* 42 (1961), 263–7.
—— 'The Autographs of John Blow', *MR* 25 (1964), 85–95.
—— 'The Harpsichord Music of John Blow: A First Catalogue', in O. W. Neighbour (ed.), *Music and Bibliography: Essays in Honour of Alec Hyatt King* (London, 1980), 51–68.
—— *The Bing–Gostling Part-Books at York Minster* (Croydon, 1986).
—— *The Succession of Organists of the Chapel Royal and the Cathedrals of England and Wales from* c.*1538* (Oxford, 1991).
SHAW, W. A., 'Three Unpublished Portraits of Henry Purcell', *MT* 61 (1920), 588–90.
—— (ed.), *Calendar of Treasury Books Preserved in the Public Record Office*, ix: *1689–1692* (London, 1931).
SHAY, R. S., 'Henry Purcell and "Ancient" Music in Restoration England' (Ph.D. diss., University of North Carolina, Chapel Hill, 1991).
SHUTE, J. D., 'Anthony à Wood and his Manuscript Wood D 19 (4) at the Bodleian Library, Oxford' (Ph.D. diss., International Institute of Advanced Studies, Clayton, Mo., 1979).
SIMPSON, C., *A Compendium of Practical Musick* (London, 1667); ed. P. J. Lord (Oxford, 1970).
SIMPSON, C. M., *The British Broadside Ballad and its Music* (New Brunswick, NJ, 1966).

SMALLMAN, B., 'Endor Revisited: English Biblical Dialogues of the Seventeenth Century', *ML* 46 (1965), 137–45.
SMITHERS, D., *The Music and History of the Baroque Trumpet before 1721* (London, 1973).
SPENCER, R., 'The Performance Style of the English Lute Ayre *c.*1600', *The Lute* (1984), 55–68.
SPINK, I., 'Angelo Notari and his "Prime Musiche Nuove"', *MMR* 87 (1957), 168–77.
—— 'Playford's "Directions for Singing after the Italian Manner"' *MMR* 89 (1959), 130–5.
—— 'The Old Jewry "Musick-Society", a Seventeenth-Century Catch Club', *Musicology*, 2 (1967), 25–41.
—— *English Song from Dowland to Purcell* (2nd edn., London, 1987).
—— 'Church Music II: From 1660', in Spink (ed.), *Seventeenth Century*, 97–137.
—— (ed.), *The Seventeenth Century* (The Blackwell History of Music in Britain, 3; Oxford, 1993).
SQUIRE, W. B., 'Purcell as Theorist', *SIMG* 6 (1904–5), 521–67.
—— 'An Unknown Autograph of Henry Purcell', *MA* 3 (1911–12), 5–17.
STERNFELD, F. W., FORTUNE, N., and OLLESON, E. (eds.), *Essays on Opera and English Music in Honour of Sir Jack Westrup* (Oxford, 1975).
STEVENS, D., 'Purcell's Art of Fantasia', *ML* 33 (1952), 341–5.
TESELLE BOAL, E., 'Purcell's Clock Tempos and the Fantasias', *JVGSA* 20 (1983), 24–39.
THOMPSON, E., 'English Biblical Dialogues', *ML* 46 (1965), 289–90.
THOMPSON, R. P., 'English Music Manuscripts and the Fine Paper Trade 1648–1688', (Ph.D. thesis, London, 1988).
—— Review of M. Campbell, *Henry Purcell: Glory of his Age*, in *Chelys*, 22 (1993), 49–50.
—— 'Purcell's Great Autographs', in Price (ed.), *Purcell Studies*.
TILMOUTH, M., 'Chamber Music in England, 1675–1720' (Ph.D. thesis, Cambridge, 1959).
—— 'The Technique and Forms of Purcell's Sonatas', *ML* 40 (1959), 109–21.
—— 'A Calendar of References to Music in Newspapers Published in London and the Provinces (1660–1719)', *RMARC* 1 (1961); *RMARC* 2 (1962), 1–15.
VAN LENNEP, W. (ed.), *The London Stage 1660–1800*, i: *1660–1700* (Carbondale, Ill., 1965).
WAINWRIGHT, J. P., 'George Jeffries' Copies of Italian Music', *RMARC*, 23 (1990), 109–24.
—— 'The Musical Patronage of Christopher, First Baron Hatton (1605–1670)' (D.Phil. thesis, Oxford, 1992).

WALKER, F. H., 'Purcell's Handwriting', *MMR* 72 (Sept. 1942), 155–7.
WALKLING, A. R., 'Politics and the Restoration Masque: The Case of *Dido and Aeneas*', in G. MacLean (ed.), *Literature, Culture, and Society in the Stuart Restoration* (Cambridge, forthcoming).
WESS, J., '*Musica Transalpina*, Parody, and the Emerging Jacobean Viol Fantasia', *Chelys*, 15 (1986), 3–25.
WESTRUP, J. A., 'Foreign Musicians in Stuart England', *MQ* 27 (1941), 70–89.
—— *Purcell* (4th edn., London, 1980).
WHITE, E. W., 'Early Theatrical Performances of Purcell's Operas', *Theatre Notebook*, 13 (1959), 2–24.
—— 'New Light on *Dido and Aeneas*', in Holst (ed.), *Purcell: Essays on his Music*, 14–34.
WILLETTS, P., 'Stephen Bing: A Forgotten Violist', *Chelys*, 18 (1989), 3–17.
WILSON, J. (ed.), *Roger North on Music* (London, 1959).
WILSON, M., *The English Chamber Organ: History and Development 1650–1850* (Oxford, 1968).
WOOD, B., 'Two Purcell Discoveries—2: A Coronation Anthem Lost and Found', *MT* 118 (1977), 466–8.
—— 'A Newly Identified Purcell Autograph', *ML* 59 (1978), 329–32.
—— 'The First Performance of Purcell's Funeral Music for Queen Mary', in Burden (ed.), *Performing the Music of Purcell*.
—— 'Only Purcell E're Shall Equal Blow', in Price (ed.), *Purcell Studies*.
ZIMMERMAN, F. B., 'Purcell and Monteverdi', *MT* 99 (1958), 368–9.
—— 'A Newly Discovered Anthem by Purcell', *MQ* 45 (1959), 302–11.
—— *Henry Purcell 1659–1695: An Analytical Catalogue of his Music* (London, 1963).
—— 'Purcell's "Service Anthem" "O God, Thou art my God", and the B flat major Service', *MQ* 50 (1964), 207–14.
—— 'Anthems of Purcell and Contemporaries in a Newly Rediscovered Gostling Manuscript"', *AcM* 41 (1969), 55–72.
—— *Henry Purcell, 1659–1695: His Life and Times* (2nd edn., Philadelphia, 1983).

Music

BANISTER I, J., and Lowe, T., *New Ayres and Dialogues* (London, 1678).
BARLOW, J. (ed.), *The Complete Country Dance Tunes from Playford's Dancing Master 1651–ca. 1728* (London, 1985).
BLOW, J., *Amphion Anglicus* (London, 1700; repr. 1965).
—— *Salvator mundi*, ed. H. W. Shaw (London, 1949).
—— *Begin the Song*, ed. H. W. Shaw (London, 1950).
—— *Chaconne for String Orchestra*, ed. H. W. Shaw (London, 1958).

—— *Ode on the Death of Mr. Henry Purcell*, ed. W. Bergmann (London, 1962).
—— *A Masque for the Entertainment of the King: Venus and Adonis*, ed. C. Bartlett (Wyton, 1984).
—— *Anthems II: Anthems with Orchestra*, ed. B. Wood (MB 50; London, 1984).
—— *O God, Wherefore Art Thou Absent from Us* (Chichester, 1990).
—— *Anthems III: Anthems with Strings*, ed. B. Wood (MB 64; London, 1993).
BOYCE, W. (ed.), *Cathedral Music* (2nd edn., London, 1788).
CARR, J., *Tripla concordia* (London, 1677).
CORBETT, W., *Six Sonatas with an Overture in 4 Parts* (London, 1708).
CROFT, W., *Te Deum*, ed. H. W. Shaw (London, 1979).
—— *The Burial Service*, ed. B. Wood (Croydon, 1985).
CUTTS, J. P. (ed.), *La Musique de scène de la troupe de Shakespeare* (Paris, 1959).
DART, T. (ed.), *The Second Part of Musick's Hand-Maid* (EKM 10; 2nd edn., London, 1968).
DOE, P. (ed.), *Elizabethan Consort Music I* (MB 44; London, 1979).
D'URFEY, T. (ed.), *Songs Compleat, Pleasant and Divertive* (London, 1719; repr. 1872).
GIBBONS, C., *Keyboard Compositions*, ed. C. G. Rayner, rev. J. Caldwell (CEKM 18; 2nd edn., Stuttgart, 1989).
GIBBONS, O., *Consort Music*, ed. J. Harper (MB 48; London, 1982).
HILLIER, P. (ed.), *The Catch Book* (Oxford, 1987).
HILTON, J., *Catch That Catch Can* (London, 1652; repr. 1970).
HUMFREY, P., *Complete Church Music I, II*, ed. P. Dennison (MB 34, 35; London, 1972).
JENKINS, J., *Consort Music of Four Parts*, ed. A. Ashbee (MB 26; London, 1969).
—— *The Lyra Viol Consorts*, ed. F. Traficante (RRMBE 67–8; Madison, Wis., 1992).
LEGRENZI, G., *Sonate a Due a Tre Opus 2 1655*, ed. S. Bonta (Harvard Publications in Music, 14; Cambridge, Mass., 1984).
LE HURAY, P. (ed.), *The Treasury of English Church Music*, ii: *1545–1650* (London, 1965).
LOCKE, M., *Melothesia* (London, 1673; repr. 1975); *Seven Pieces (Voluntaries) from 'Melothesia' (1673)*, ed. G. Phillips (Tallis to Wesley, 6; London, 1957).
—— *Chamber Music I, II*, ed. M. Tilmouth (MB 31, 32; London, 1971, 1972).
—— *Anthems and Motets*, ed. P. le Huray (MB 38; London, 1976).
—— *Suite in G from Tripla Concordia*, ed. P. Holman (London, 1980).
—— *Dramatic Music*, ed. M. Tilmouth (MB 51; London, 1986).

LOCKE, M., *The Rare Theatrical, New York Public Library, Drexel MS 3976*, ed. P. Holman (MLE A-4; London, 1989).
—— and GIBBONS, C., *Cupid and Death*, ed. E. J. Dent (MB 2; 2nd edn., London, 1965).
MATTEIS, N., *Concerto in C*, ed. P. Holman (London, 1982).
NYMAN, M. (ed.), *Come Let us Drink* (Great Yarmouth, 1972).
PACHELBEL, J., *Canon and Gigue*, ed. C. Bartlett (Wyton, 1990).
PIGNANI, G., *Scelta di canzonette Italiane de più autori* (London, 1679).
PLAYFORD, H., *The Theater of Music* (1685–7); ed. R. Spencer (MLE A-1; Tunbridge Wells, 1983).
—— *Apollo's Banquet* (5th edn., London, 1687; 8th edn., 1701).
—— *The Banquet of Musick*, i, ii (London, 1688).
—— *Harmonia sacra*, i (London, 1688; 2nd edn., 1703; repr. 1714, 1726, 1966); ii (London, 1693; 2nd edn., 1714; repr. 1726, 1966).
PLAYFORD, J., *Catch That Catch Can; or, The Musical Companion* (London, 1667).
—— *The Treasury of Musick* (London, 1669; repr. 1966).
—— *Psalms and Hymns in Solemn Musick* (London, 1671).
—— *The Musical Companion* (London, 1673).
—— *Cantica sacra . . . the Second Sett* (London, 1674).
—— *Choice Ayres, Songs and Dialogues* (London, 1673–84); ed. I. Spink (MLE A-5; London, 1989).
PRICE, C. A. (ed.), *Don Quixote: The Music in the Three Plays of Thomas Durfey* (MLE A-2; Tunbridge Wells, 1984).
PORTER, W., *Madrigales and Ayres* (London, 1632; repr. 1970).
PURCELL, H., *The Works of Henry Purcell* [*Works I, II*] (London, 1878–).
I, 1 *The Yorkshire Feast Song*, ed. W. H. Cummings (1878).
II, 2 *Timon of Athens*, ed. F. A. Gore Ouseley, rev. J. A. Westrup (1974).
II, 3 *Dido and Aeneas*, ed. A. M. Laurie (1979).
II, 4 *A Song for the Duke of Gloucester's Birthday (1695)*, ed. I. Spink (1990).
II, 5 *Twelve Sonatas in Three Parts*, ed. M. Tilmouth (1976).
I, 6 *Harpsichord and Organ Music*, ed. W. B. Squire and E. J. Hopkins (1895).
II, 7 *Ten Sonatas in Four Parts*, ed. M. Tilmouth (1981).
II, 8 *Ode on St Cecilia's Day 1692*, ed. P. Dennison (1978).
II, 9 *Dioclesian*, ed. J. F. Bridge and J. Pointer, rev. A. M. Laurie (1961).
II, 10 *Three Odes for St Cecilia's Day*, ed. B. Wood (1990).
II, 11 *Birthday Odes for Queen Mary, Part 1*, ed. B. Wood (1993).
II, 12 *The Fairy Queen*, ed. J. S. Shedlock, rev. A. Lewis (1968).
II, 13 *Sacred Music, Part I: Nine Anthems with Orchestral Accompaniment*, ed. P. Dennison (1988).

II, 14 *Sacred Music, Part 2*, ed. P. Dennison (1973).
I, 15 *Welcome Songs I*, ed. R. Vaughan Williams (1905).
I, 16 *Dramatic Music I*, ed. A. Gray (1906).
II, 17 *Sacred Music, Part III: Seven Anthems with Strings*, ed. H. E. Wooldridge and G. E. P. Arkwright, rev. N. Fortune (1964).
I, 18 *Welcome Songs II*, ed. R. Vaughan Williams (1910).
I, 19 *The Indian Queen and The Tempest*, ed. E. J. Dent (1912).
I, 20 *Dramatic Music II*, ed. A. Gray (1916).
I, 21 *Dramatic Music III*, ed. A. Gray (1917).
I, 22 *Catches, Rounds, Two-part, and Three-part Songs*, ed. W. B. Squire and J. A. Fuller-Maitland (1922).
I, 23 *Services*, ed. A. Gray (1923).
I, 24 *Birthday Odes for Queen Mary II*, ed. G. Shaw (1926).
II, 25 *Secular Songs for Solo Voice*, ed. A. M. Laurie (1985).
II, 26 *King Arthur*, ed. D. Arundell, rev. A. M. Laurie (1971).
I, 27 *Miscellaneous Odes and Cantatas*, ed. A. Goldsbrough, D. Arundell, A. Lewis, and T. Dart (1957).
I, 28 *Sacred Music, Part IV: Anthems*, ed. A. Lewis and N. Fortune (1959).
I, 29 *Sacred Music, Part V: Anthems*, ed. A. Lewis and N. Fortune (1960).
I, 30 *Sacred Music, Part VI: Songs and Vocal Ensemble Music*, ed. A. Lewis and N. Fortune (1965).
II, 31 *Fantazias and Miscellaneous Instrumental Music*, ed. T. Dart, rev. M. Tilmouth, A. Browning, and P. Holman (1990).
I, 32 *Sacred Music, Part VII: Anthems and Miscellaneous Church Music*, ed. A. Lewis and N. Fortune (1962).

—— *Sonnata's of III Parts* (London, 1683; repr. 1975).

—— *Ten Sonata's in Four Parts* (London, 1697; repr. n.d.).

—— *A Collection of Ayres, Compos'd for the Theatre, and upon other Occasions* (London, 1697).

—— *Orpheus Britannicus*, i (London, 1698; 2nd edn., 1706; repr. 1721, 1965), ii (London, 1702; 2nd edn., 1706, repr. 1721, 1965).

—— *Come ye Sons of Art*, ed. M. Tippett and W. Bergmann (London, 1951).

—— *Fantasia Three Parts upon a Ground*, ed. D. Stevens and T. Dart (London, 1953).

—— *Ode for St. Cecilia's Day (1692)*, ed. M. Tippett and W. Bergmann (London, 1955).

—— *Dido and Aeneas*, ed. T. Dart and A. M. Laurie (London, 1961).

—— *Welcome to all the Pleasures, Ode for St. Cecilia's Day 1683*, ed. W. Bergmann (London, 1964).

—— *Organ Works*, ed. H. McLean (2nd edn., London, 1967).

PURCELL, H., *Complete Harpsichord Works*, ed. H. Ferguson (EKM 21, 22; 2nd edn., London, 1968).
—— *I Was Glad when They Said unto Me*, ed. B. Wood (London, 1977).
—— *Dido and Aeneas*, ed. C. A. Price (New York and London, 1986).
—— *Dido and Aeneas*, ed. E. Harris, full score (Oxford, 1987).
RAMSEY, R., *English Sacred Music*, ed. E. Thompson (EECM 7; London, 1967).
REGGIO, P., *Songs Set by Pietro Reggio* (London, 1680).
ROBERTS, T. (ed.), *Bring Me Poison, Daggers, Fire: Thirteen Songs of Passion and Madness by Henry Purcell and his Contemporaries* (Oxford, forthcoming).
SABOL, A. (ed.), *Four Hundred Songs and Dances from the Stuart Masque* (Providence, 1978).
SCHERING, A. (ed.), *Geschichte der Musik in Beispielen* (Leipzig, 1931).
SCHMELZER, J. H., *Sacro-profanus concentus musicus* (Nuremberg, 1662); ed. E. Schenk (DTÖ 111/112; Graz and Vienna, 1965).
SPINK, I. (ed.), *English Songs 1625–1660* (MB 33; London, 1971).
WILSON, J., *Cheerful Ayres* (Oxford, 1660).
YOUNG, W., *Sonate à 3. 4. e 5* (Innsbruck, 1653); ed. H. and O. Wessely (DTÖ 135; Graz, 1983).
ZIMMERMAN, F. B. (ed.), *The Gostling Manuscript*, facs. (Austin, Tex., and London, 1977).

INDEX OF PURCELL'S WORKS

Theatre works

Keyboard music

GENERAL INDEX